Adaptations in the Animal Kingdom

Adaptations in the Animal Kingdom

Verne A. Simon

To order additional copies of this book, contact:
Xlibris Corporation
1-888-795-4274
www.Xlibris.com
Orders@Xlibris.com
71287

Contents

Preface

The purpose of this book is to entertain and to educate. It is intended for both a general audience as well as scholars of animal science. The breadth of topics is limited in keeping with the modest length of this small volume. Many quite worthy aspects of animal science and behavior are not covered. My guiding principle is to try to make the text insightful and interesting. The first portion of the book discusses general principles concerning adaptations and strategies for survival. It is hoped that this part of the book may bring some thought-provoking considerations to the reader's attention.

The second part of the book concerns a somewhat detailed examination of selected species. The selection is quite limited and arbitrary. The animals chosen have been selected because there is something interesting and novel to be said about them. An eclectic approach is intended to make the book more entertaining and subject matter more memorable.

Writing the book has been an intellectual adventure for me, and I hope that readers will share in the joy of contemplation of our fellow animals that share our world. You may even think of animals in a different way after reading this book.

Chapter 1

Temperature Regulation in Animals

There are, broadly speaking, two kinds of animals with regard to body temperature: exothermic (cold-blooded) and endothermic (warm-blooded) animals. The exothermic animals, such as reptiles, do not supply body heat by metabolic conversion of food to heat. Reptiles allow their surroundings to determine their body temperature. They lie out in the sun to warm their body. If they are too hot, they seek the shade or even burrow into the ground. At night they hide from the cold in burrows or squeeze into cracks between rocks or hide in leaf cover. Reptiles avoid the extremes of temperature. When reptiles become cool, their movements slow down, and chemical processes in their bodies, such as digestion, are inhibited. Predators, such as hawks and eagles, find it easier to prey on lizards and snakes in cooler weather. The distribution of reptiles is somewhat limited by their exothermic character. They do not thrive in cold climates[1].

What are the advantages and disadvantages in being exothermic? When the lizard is in a cool environment and cannot find a warmer spot, its body simply cools to the temperature of the surroundings. It is not necessary for the exothermic lizard to generate heat to increase its body temperature. This means that the lizard uses less energy and does not have to eat as much. As the lizard cools its digestion, breathing rate and heart rate slow, saving energy. A disadvantage occurs when the cool lizard is attacked by a predator. If warm, he could run fast and have a much better chance of

---

[1] St. Patrick did not chase the snakes out of Ireland. Ireland was already completely free of snakes. St. Patrick was instrumental in converting pagans to Christianity. Since the snake was a symbol used in pagan rituals, St. Patrick was influential in ridding Ireland of the ritual use of symbolic snakes.

evading capture. A warm lizard being chased by a predator can move quite fast for a short distance, but like other exotherms, lacks endurance and soon tires. When the exotherm is running fast, its effort is anaerobic, that is, is not using oxygen, and lactic acid is building up in its body. It soon tires and is unable to exert itself. It must recover by taking in oxygen to rid the body of lactic acid. Another disadvantage of exothermic life is that cold climates are not available as habitat. If there is a sudden climate change, an exothermic animal wouldn't be able to mount the sustained effort needed to migrate to a better environment. The exothermic creature might simply perish.

About 180 million years ago, mammals appeared. Mammals are endothermic (warm-blooded) and are able to maintain a nearly constant body temperature regardless of the temperature of their surroundings within wide limits. Their bodies will not tolerate too high or too low a temperature. If the surroundings are too hot or cold, causing the body temperature to exceed allowed limits, the animal will die. Mammals have furry coats to help them tolerate low temperatures. Sea-dwelling mammals—whales, seals, and walrus—have thick layers of blubber for insulation. Birds are endothermic and have feathers to protect them from the cold. Many types of birds and mammals survive in cold climates. Emperor penguins even live in the Antarctic, in the coldest climate on earth. Under normal circumstances, mammals and birds manage to keep this very nearly constant body temperature regardless of the temperature of their surroundings. Mammals are characterized by having body hair and suckling their young. This latter behavior gives the class its name; mammals must have mammary glands. A second advantage is that endothermic animals are not limited to activity only in daylight hours. In many locations, it is too cold at night for exotherms to be active. Even very cold temperatures do not exclude endothermic animals such as mammals and birds from nocturnal activity. Exothermic animals are not normally found in cold climates, though there are a few exceptions. Mammals with their hair can grow warm fur coats as in the polar bear or beaver. Mammals with little or no hair often have a thick subcutaneous layer of fat for thermal insulation as in walrus and whales. Birds have feathers for insulation, like the snowy owl, whose white feathers match the snow.

During long periods of cold weather (winter), some, but not all, mammals hibernate. Their body temperature drops to a few degrees above freezing, and their breathing and heart rates almost cease. Other mammals, such as bears, undergo estivation. Their body temperature may drop by 40°F (22°C). While in estivation, they live off the fat they have accumulated in preparation for winter. They may become active for short periods on relatively warm days. Some animals tough it out in winter, like emperor penguins and moose.

Mammals in cold latitudes develop furry coats to retain heat during cold winters. Mammals in warm climates may have very sparse hair, for example, elephants and naked mole rats. Some mammals (sheep, horses, cattle) give birth to single young with well-developed senses of sight and hearing and with musculature ready for walking within minutes of birth. Such well-developed young are said to be precocial.

In contrast to precocial young are altricial young, born in litters, with closed eyes, closed ear canals, having no capability of locomotion, and often sparsely covered with hair or feathers. Some examples of endothermic altricial young are dogs, cats, rats, mice, and polar bears. Human babies are neither clearly altricial nor precocial. At birth a human baby's eyes are open and the baby can hear. On the other hand, a human infant is naked and unable to move about. Humans usually are born singly. Altricial animals are usually born in litters and are small. Because of their small size they have a large surface to volume ratio, which makes it more difficult to maintain a constant body temperature as required by endothermic animals.

How do altricial young cope with temperature fluctuations? Some animals such as puppies and kittens have numerous litter mates with which they can huddle, thereby decreasing their exposed surface area. Mothers share bodily heat with their offspring and provide warm milk as a buffer against the cold. There is a special organ for heat production in young altricial animals, that is, brown adipose tissue (BAT). Heat production by BAT is stimulated by cold temperatures and by norepinephrine[2] (noradrenaline). BAT is present in large amounts in the young, constituting about 5% of body weight. In adults, BAT constitutes only 1% of body weight and has traditionally been regarded as unimportant. Adults are able to raise their body temperature by the muscular activity, including shivering, but infants seem unable to shiver. Human babies have BAT deposits in the neck and between the shoulder blades. Heat production by BAT in human babies can be demonstrated by infrared photography. An unclothed baby lying prone at room temperature shows hot spots between his shoulder blades and in the area of his neck in an infrared photo. The distribution of BAT has important advantages. It is very helpful in keeping the heart and brain from cooling. Both of these organs show decreased activity if the temperature drops. The placement of BAT on neck and shoulders aids in keeping the brain and heart warm.

BAT is brown due to pigmented organelles called mitochondria within this special tissue. The BAT is very rich in mitochondria. Mitochondria are

---

[2] Epinephrine (adrenaline) is a hormone produced by glands at the top of the kidneys. Epi means "above" and nephrine refers to the kidney. Norepinephrine is an active metabolite of epinephrine. Both epinephrine and norepinephrine prepare animals for "fight or flight."

present in most body cells and have the function of producing energy which can be chemically stored in the form of adenosine triphosphate (ATP). Through muscular activity, this stored energy is converted into mechanical energy with heat production as a by-product. In BAT, the mitochondria have been modified with a brown material which allows the energy to be immediately released as heat rather than stored as ATP. Ordinary fat or white adipose tissue (sometimes called WAT) can be used as fuel for energy. BAT is also important for warming adult animals awakening from a state of hibernation. Animals recovering from hibernation are too cold to be able to shiver. The chemical processes which instigate shivering are too slow when the animal is first awakening from hibernation.

Marsupials produce young so altricial that they are in a class by themselves. Marsupials by definition produce newborns at such an early stage of development that the young are very small (about the size of a jelly bean). To protect such fragile babies, the mother usually, but not always, possesses a pouch for carrying the young. Some marsupials, like the numbat, do not have a pouch. Marsupial newborns attach themselves to their mother's teat, which swells and so fills the infant's mouth that the baby cannot detach itself. Examples of marsupials with a pouch are kangaroos, koalas, and possums. The Virginia opossum is the only marsupial that lives in the United States. Marsupials do not have a placenta needed to nourish the fetus through a long gestational period. They have a yolk sac which provides nourishment to the fetus, but only long enough to reach a very early stage of development. When this yolk sac is depleted, the baby or babies are born while they are very small. Marsupial newborns are not able to suck. Muscles in the mother's teat contract, ejecting milk to feed the baby. Obviously, these jelly-bean-sized young have a very large surface to volume ratio and would be subject to danger by heat loss. They are protected by being constantly in intimate contact with their mother[3]. A pouch on their mother's belly is called a marsupium. Marsupium is a Latin word for a purse used in Ancient Rome for carrying money[4].

Polar bear cubs present a particularly challenging case of thermal threat. The polar bear usually gives birth to altricial twins or singletons in December, the coldest time of the year. The temperature outside her snow den is -40°F (-40°C, not an error, just a coincidence). Inside the den, the temperature is around 32°F (0°C). The cub or cubs weigh only about 2.2 lb. (1 kg) and are nearly naked. They are blind and have little subcutaneous fat for insulation. They have very little capacity to produce heat. The mother gives birth while

---

[3]   It has been suggested that human babies would benefit from closer bodily contact with their mothers. Close frequent contact with the mother is sometimes called *kangaroo* care.

[4]   Marsupials and song birds lack a corpus callosum, which connects the right and left brains.

she is estivating and sleeps through the entire process, but she curls up with the cub/cubs and provides them with warmth. She provides them with milk of high fat content. In three months, the cubs emerge from the den, weighing ten times their birth weight. They are completely furred and able to maintain a constant body temperature in spite of a very low air temperature of -22°F (-30°C).

Although it has, until recently, been assumed that BAT has little significance for adult humans, there is now known to be a correlation between BAT content and a slender body shape. BAT is found in adults, especially in women, mainly around the anterior neck and upper thorax. Deposits of BAT can be detected by positron-emission tomography. Already, there have been attempts to promote clothing with holes under the arms and in the back, which claim to sculpt the body by promoting BAT formation. Not surprisingly, the commercial claims far outstrip any scientific support. Scientists have been able to stimulate BAT production in rats, but similar tests have not yet been carried out on humans.

Body organs may require special individual temperature control. There is a need to fine-tune the temperature of certain organs. A case in point is temperature regulation of the testes. In the early life of a human male fetus, the testes are located in the abdominal cavity. Normally the testes descend by about the eighth month of gestation through the inguinal canal into the scrotum. Sometimes the testes fail to descend and remain in the abdomen, or descend only partially, lodging in the inguinal canal. This condition is known as cryptorchidism and results in sterility. Testes in the scrotum are 4-16°F (2-9°C) cooler than body temperature. This cooling is due partially to the usually cooler surrounding air. Cooling of the testes is aided by another strategy, countercurrent heat exchange. Warm arterial blood from the heart on its way to the testes passes in proximity to the cooler venous blood coming from the testes on its way back to the heart. This precools the blood supplied to the testes. The body has another mechanism to protect male reproduction. In the scrotum are dartos muscle fibers which contract pulling the scrotum and testes closer to the body when subjected to cold temperatures thereby warming the testes. When warmed, the dartos relaxes, and the scrotum hangs lower, cooling the testes. Men who would like to be fathers are ill advised to wear bikini underwear as this brings the testes close to the body and inhibits viable sperm production.

Chapter 2

Considerations of Scale

The surface area of an animal relative to its weight or volume has a fundamental influence on its lifestyle and metabolism. Very small animals such as mice, shrews, and hummingbirds have large surface to weight ratios. This means that on a cool day, heat can escape easily from a hummingbird's body. Heat can only leave the animal at its surface. Even at the center of the bird's body, heat doesn't have far to go to reach the surface. If the hummingbird wants to stay warm, it needs to eat a lot of nectar, its own weight in nectar each day. In cool air, a hummingbird must metabolize its food at a great rate in order to compensate for heat loss.

An endothermic animal must avoid overheating as well as cooling. In order to gather nectar, a hummingbird must execute very vigorous flight. Vigorous activity causes a rise in body temperature. A medium-sized species of hummingbird beats its wings twenty to twenty-five times per second. Hummingbirds cannot maintain flight constantly. In fact, 75-80% of the time, the hummingbird rests. A resting hummingbird's pulse rate is 250 beats per minute. The rate increases to 1,200 beats per minute when in flight. Muscular activity increases body temperature, so the bird frequently rests in the shade to avoid overheating.

Hummingbirds cannot feed in the dark, and so their metabolism has to be slowed during the night; otherwise, they would starve to death before morning. Every night, the hummingbird goes into a state of torpor (something akin to mild hibernation). It is possible to pick up and handle a sleeping hummingbird. If a hummingbird is losing heat, it doesn't take very much loss to change the bird's temperature. Hummingbirds are always a few hours away from starvation. Hummingbirds consume 100% of their own weight in nectar each day. This percentage would be unheard-of for large animals. Nectar does not supply vitamins and minerals. The bird gets these by eating insects and spiders. Ruby-throated hummingbirds migrate nonstop across the Gulf of Mexico. To acquire the necessary energy for the trip, the bird doubles its weight by fattening up before migration. Hummingbirds have a relatively short life span. Many do not make it through their first year. On average, hummingbirds live only three or four years. A high rate of metabolism is often thought to correlate with short life span. This correlation is not to be relied on, as most birds live much longer than would be expected based on metabolic rates.

With an elephant, the situation is quite different. There is heat generated inside the elephant's intestines from bacteria fermenting its food. The tropical sun adds more heat to the elephant's body. This heat may slowly leave the elephant, but it takes a long time. The 100 gallons of warm blood is circulated through the elephant's ears every twenty minutes. The elephant's ears have a large surface area and are quite thin. Since the ears are so thin, heat is quickly transferred to the surrounding air. A further adaptation for heat loss is the elephant's wrinkled skin, which increases its surface area aiding in heat transfer. An overly warm elephant is able to relax muscles around its blood vessels, encouraging increased blood flow to the ears. The elephant can also increase the cooling effect by flopping its ears. The elephant's ears are like the radiator of a car. Flopping its ears is like the radiator fan increasing heat transfer to the air. To cool an elephant by a single degree, you must cool down 14,000 lb. of flesh. This involves transfer of a lot of heat, and it cannot be accomplished rapidly. The mere bulk of a large animal helps to smooth out body temperature fluctuations. Small animals are much more vulnerable to the deleterious effects of temperature extremes than are large animals.

Elephants rest in the shade during the hottest part of the day. Elephants remain standing and are not asleep in midday. (See chapter 6 to find out why.) In hot weather, the elephant loves to wallow in water or mud. Water conducts heat much faster than air does and has a higher capacity to hold heat (heat capacity) than air. Cooling off in a river or lake is more effective than standing in the shade. Many animals, but not elephants or hummingbirds, sweat when overheated. Evaporation of sweat can help in cooling the body. It takes energy to change liquid water into vapor, and in a sweating animal, such as a horse or a human, this energy, in the form of heat, is withdrawn from the body.

If an elephant had the metabolic rate of a hummingbird, it would be thoroughly cooked in short order because the extra heat could not escape from its body. Obviously, an elephant's metabolic rate per unit body weight is much lower than that of a bird's. The pulse rate of an elephant is 28. Even though an elephant's metabolic rate is low, it still must consume large quantities of food because its herbivorous diet is not very nutritious and its digestive system is not very efficient. The elephant is a hindgut digester.

Also, there is a lot of elephant tissue to satisfy. Some processes of metabolism requiring energy must continue just to keep its extensive tissues alive.

It is ironic that the hummingbird must eat prodigious amounts of food because it is so small and the elephant must eat so much because it is so large. Besides being large, the elephant doesn't process its food efficiently, and its food is not very nutritious anyway. Part of the reason the elephant is so large is that it needs to be able to travel great distances to gather enough food. Is the elephant large in order to gather enough food, or is the elephant so large because it eats so much? At any rate, one could say that elephants eat grass because the grass is there to be eaten. To put it another way, the savanna provides the elephant with an ecological niche. Since there are only so many niches available, to survive, animals had better make use of their opportunities. This tendency to fill ecological niches has surely led to a profusion of animal and plant species. It has also led to very frequent extinctions when niches disappear. It is not generally appreciated how frequently extinctions occur in nature. Fully 99% of all animals that have ever lived are extinct. In a sense, many of these animals are still with us as their genes have been handed down to animals which evolved from those that went extinct.

Certain hummingbirds have carved out an ecological niche by acquiring the ability to gather nectar from long tubular flowers. These hummingbirds have bills and tongues particularly suited for pollinating flowers that cannot be pollinated by other birds or insects. It appears that the unusual bills of the sword-billed hummingbird and the sicklebills coevolved with a small number of flower species. That is, without the particular flowers, there wouldn't ever have been sword-bills or sicklebills, and, conversely, without the sword-bills and sicklebills, these flowers would never have evolved. Another trick the hummingbird has used to carve its niche is that it has developed the ability to see farther into the ultraviolet light than do other birds and insects. Other species, for example, honeybees and other birds, can also see into the ultraviolet (UV), but not as far into the UV as hummingbirds can. This may give the hummingbirds an advantage in locating flowers which reflect UV light[5].

Very tall animals have a unique problem related to their height. In their lower limbs, the blood in their veins and arteries is under pressure from the overlying column of blood. In elephants and giraffes, this requires special adaptations to prevent blood from leaking out of these low-lying blood

---

[5]   It has been reported that some human patients, after cataract surgery, can see into the near UV and detect patterns in flowers previously not visible before surgery. I have personally talked to at least two who declared, in two independent conversations, that after this surgery they were astounded at the colors they could see.

vessels. The same problem faces a giraffe's head as he lowers it to drink. In both of these animals, there are thick layers of skin or muscle to reinforce the blood vessels. The height of a male giraffe is 15 ft., making it the tallest of all animals. Bull elephants with their long trunks can reach as high as any giraffe can. The elephant has some unique features of its circulation. The ventricular apex shows a separation. There are also two venae cavae rather than the usual single vena cava which returns blood to the heart. The hemoglobin of elephants has a higher affinity for oxygen than most animals. This is believed to be related to their size.

Small animals like mice and rats benefit by their small size, being able to hide from predators more effectively. They also require less food and are less likely to starve. Mice and rats have a partner who feeds them pretty well—man. Wherever you find people, you find rats and mice. Mice and rats are, indeed, common prey animals, but they have advantages and disadvantages, which leads to a balance. Mice and rats are not threatened species. Their secret weapon for survival is their rapid rate of reproduction. It doesn't take a long gestation period to produce a litter of a dozen mice, just twenty-one days. A female mouse can have ten litters a year. Elephants require twenty-two months to produce a single baby. To make matters worse (especially for bull elephants), female elephants are ready to mate for only two days out of every four years. Copulation is over in a few seconds. Nature can be cruel.

Human activities such as transportation, real estate development, agriculture, and filling in of wetlands have caused decimation of many animal populations. Loss of habitat is especially damaging to larger species because large animals need a bigger territory. Large animals such as big cats, elephants, hippopotamuses, rhinoceroses, and giraffes, to name a few, are threatened.

Carnivores, because of their position on the food chain, are very sensitive to anything that reduces the population of their prey. The harm done by carnivores to prey species is minimized by several factors. Some carnivores such as African wild dogs and hyenas tend to prey on the young or on old weak herd animals. The young which are taken have a reduced statistical likelihood of reproduction due to their high mortality from causes other than predation. The old or sick have very little probability of reproduction. Some predators, lions, are more likely to take healthy zebras.

Insular dwarfism will serve to illustrate the effect of competition for food among large animals. When large animals find themselves isolated from others of their kind, as when they are living on an island, there may be keen competition for food. Either by the slow process of natural selection, or by the sudden accidental birth of dwarf animals, smaller individuals occur in the population. These smaller animals can survive on the limited supply of food available, while the larger ones may die before they can reproduce. This has led to tiny elephants on the Mediterranean islands of Crete, Malta, Cyprus, and Sicily. For unknown reasons, these tiny elephants have become extinct. Dwarf raccoons still live on Cozumel Island off the Yucatán Peninsula. A dwarf wolf population developed in Japan, but was wiped out by rabies in the 1700s. Dwarf tigers lived on the island of Bali until they went extinct due to hunting and loss of habitat. Dwarf hippos exist in Africa. They weigh 400-600 lb., compared to the usual 1.5-3 tons. Dwarf mammoth fossils have been found on the Channel Islands off California, Saint Paul Island off Alaska, and Wrangel Island north of Siberia. The sauropod dinosaur, normally 80 ft. (24 m) in length, was, at one time, the world's largest land animal. The dwarf form was found in Gasler, Germany, in an area which was once an island. The dwarf sauropod was 20 ft. (6 m) long. A tiny human skeleton found in a cave on Flores, east of Bali, was considered to be a separate species and was given the name *Homo floresienses*.

Chapter 3

Nutritional Adaptations

About 25 million years ago, the weather became drier, and some forests gave way to extensive grassy plains. This new environment provided an opportunity for animals to evolve structures and behaviors suited to life on the grassy planes. Herbivores that "grazed" on grass or "browsed" on leaves of trees and shrubs faced important adaptations connected with diet. Other adaptations such as predator-prey relationships that also modified habits in major ways will be treated later.

Herbivores faced important dietary challenges—gathering and masticating tuff grass or leaves, and digesting what they had eaten. This food was difficult to chew, leading to dental challenges. Grass and leaves are difficult to chew and very difficult to digest. Furthermore, grass and leaves are a very poor source of nutrition. The most successful of large (over 10 kg/22 lb.) herbivores are the ruminants. A further problem with browsing is that leaves are often somewhat toxic.

Ruminants consist of some 155 species, including cattle, goats, sheep, bison, deer, caribou, moose, and giraffes (the largest of all the ruminants). All of these animals have four-chambered stomachs, and all ruminate (chew their cud). Once the food is gathered and swallowed, it passes down the esophagus to the first of four stomachs, the rumen, which connects seamlessly with the second stomach, the reticulum. Food passes freely between the rumen and reticulum. The reticulum has a different wall structure from the rumen.

In the rumen, the food, let's say grass, is mixed with a large amount of saliva. An adult cow produces from 30 to 35 gallons of saliva each day. This permits good mixing due to rhythmic stomach contractions. A major part of digestion occurs in the rumen. All plant cell walls are made of cellulose. Cellulose is a polymer of glucose; it is made up of thousands of glucose molecules chemically bound together in a long chain. Starch is also a polymer of glucose. The geometry of the linkage between glucose units is different in cellulose and starch. This seemingly simple difference makes starch digestible, while cellulose cannot be digested by the enzymes produced by animals. Bacteria come to the rescue for herbivores. The bacterium present in huge amounts in the rumen has an enzyme called cellulase which breaks down cellulose. The process by which bacteria digest cellulose is called fermentation. The cow's rumen is a large fermentation vat. If not for bacteria, ruminants could not digest grass or foliage. The neutral pH (a measure of acidity) of the rumen, adjusted by the presence of saliva, is ideal for the breakdown of cellulose by the bacterial enzyme, cellulase. The relationship between the ruminant and the bacterium is usually said to be symbiotic, though some biologists prefer the adjective mutual.

Besides carbon dioxide gas, much larger quantities of methane gas are produced in the rumen. Astounding quantities of gases are produced, 600 liters of methane a day on average for a cow. If something interferes with expulsion of gas, the cow would be in danger of dying from asphyxiation, being unable to breathe. Methane gas is about twenty times more efficient as a "greenhouse gas" than carbon dioxide[6]. As there are about 1.3 billion cows around the world, this is a major concern regarding global warming, and research is going on to alleviate the problem. Sheep also produce methane.

The bacteria produce all twenty-one of the essential amino acids needed by the cow to produce protein to build tissue and carry out many bodily functions. For example, all enzymes are proteins. Because all twenty-one of the amino acids needed to make protein are supplied by bacterial action, the cow does not require protein in her diet. The bacteria provide another service; they produce vitamins, especially B12. This is the sole source of B12 normally available to the cow.

After some fermentation in the rumen, a resting cow regurgitates some of the food and expels liquid portion with her tongue back into her esophagus and stomach. Then she ruminates, that is, she "chews her cud." This action defines the group of animals known as ruminants. When we are deep in thought, we like to say we are ruminating. A cow appears to be thoughtful while chewing cud[7]. It is mainly during rumination that the cow expels

---

[6] The kangaroo, an herbivore, does not produce methane. The bacteria in the kangaroo stomach are being studied in order to learn how they might be grown in the cow's stomach. Another major source of methane is termites. Huge termite mounds dot the landscape in Africa.

[7] A gum-chewing girl and a cud-chewing cow
Are somewhat alike, yet they differ somehow.
Oh, I see it now! . . .
It's the thoughtful look on the face of the cow.

methane gas by belching. Interestingly the gas is first taken into the lungs before being belched. Contrary to popular belief, most of the methane leaves the cephalic end of the cow, not the caudal end.

Finally, the contents are transferred to the cecum, colon, and rectum. These are collectively called the large intestine. Here a little digestion still takes place and more water is absorbed to dry out the feces. About 50% of the feces[8] are made up of bacteria. Digestion in ruminants is far superior, in terms of extracting maximum benefit for tissue building and energy production, to that of other herbivores.

There are several herbivores which have some variation of the ruminant digestive system. The hippopotamus has a three-chambered stomach but does not ruminate. Hippos are sometimes referred to as pseudo ruminants. The hippopotamus spends the daytime in a river but moves onto land to graze on grass every night. Hippos are territorial in the water, but not on land. Camelids (camels, alpacas, llamas, and guanacos) do ruminate. Camelids have a three-chambered stomach, and they, like the hippo, are also called pseudoruminants even though they ruminate. Peccaries (a kind of wild hog) have a four-chambered saculated stomach but do not ruminate.

Monogastric herbivores, as the name implies, have but one stomach. Digestion of cellulose is by bacterial fermentation, just as in ruminants, but it occurs in the intestines, not the stomach. For this reason some like to call them hindgut fermenters. By symmetry, ruminants are called foregut fermenters. It should be emphasized that this discussion is not about monogastric animals in general, but only monogastric herbivores. Examples of such herbivores include horses, tapirs, koala, kangaroos, rabbits, pandas, elephants, and at least one bird, the hoatzin.

Digestion in monogastric herbivores, with few exceptions (see kangaroos below), is much less efficient than in ruminants. The passage of food (digesta) through the alimentary canal is usually faster in monogastric herbivores than in ruminants, giving less time for digestion to occur. They achieve a smaller percentage of digestion to a degree which permits absorption of nutrients into the bloodstream. It has been estimated that the elephant digests only half of its food intake.

Let us consider the domestic horse as an example of a monogastric herbivore. Hay, grain, or grass enters the mouth, is chewed and passed via the esophagus to the simple stomach. It is not returned to the mouth for rumination. The stomach of a horse is much smaller than that of a cow. Horses are often fed grain to supplement grass or hay. Digestion of grain and protein occurs in the stomach and small intestine by enzymes produced

---

[8]   The word feces has no singular in English. It is an example of a plurale tantum. Other pluralia tantum are scissors, pants, and shears.

by the horse. Because the horse's stomach is small, the horse should be fed frequent small meals, and overfeeding must be avoided. Stomach emptying is regulated by food intake. Larger meals lead to rapid passage through the stomach and less complete digestion. If the horse's stomach remains empty for too long, excessive gas accumulates. The horse is incapable of either belching or vomiting, and excessive gas can actually kill the horse. Because of inability to vomit, the horse is more susceptible to poisoning. Many common plants in pastureland contain toxins. The digesta is passed from the stomach to the small intestine where additional digestion of starches and protein occurs. The esophagus, stomach, and small intestine are called the foregut.

The digesta next passes to the hindgut consisting of the cecum, the large intestine, the colon, and finally the rectum. In large monogastric herbivores (more than 22 lb./10 kg), digestion of cellulose occurs in the large intestine, with some occurring in the cecum. As with ruminants, the horse has to depend on bacteria to digest cellulose by fermentation. The horse does not have the fermentation vat with copious saliva and good mixing as does the cow. The horse does not ruminate. This leaves the digesta to be in larger coarse particles. Larger particle size and poor mixing mean less surface area on which cellulase can act. It is only at the surface of the digesta that contact with digestive enzymes can occur. Horse feces still contain a lot of the food value that the horse failed to get.

An interesting variation on monogastric herbivorous digestion can be seen in the kangaroo. In the kangaroo, fermentation is carried out in a long tubiform foregut preceding the part of the stomach in which secretion of hydrochloric acid occurs. Fermentation cannot occur in an acidic medium. Fermentation in this foregut is similar to the first stage of digestion in the cow before regurgitation and cud chewing. The kangaroo qualifies as a foregut digester. In addition, the kangaroo, though not normally referred to as a ruminant, does chew cud to some extent. Though the kangaroo does not have a multichambered stomach, it has the advantages shared by ruminants like the cow. Digestive efficiency is higher in the kangaroo than in hindgut digesters like the horse. Interestingly, kangaroos' digestion does not produce methane.

Omnivores are animals that eat a wide variety of foods. Examples are humans, domestic pigs, and rats. Men and rats eat everything. Omnivores generally have simple stomachs. Omnivores are equipped with digestive enzymes in the mouth, and small intestines which adequately digest their food. Hydrochloric acid in the stomach aids digestion. The omnivorous diet is the antithesis of a specialized diet and allows for adaptation to limitations on food supply.

An important part of nutrition involves mastication, and so the next topic concerns teeth. Rodents—such as rats, mice, squirrels, and beavers—have incisor teeth with open roots which permit the teeth to go grow throughout the animal's lifetime. The new growth is continually worn away by constant gnawing of coarse materials. The word rodent comes from the Latin verb rodere, meaning "to gnaw." Rodents have the dental formula I1/1, C0/0, P0/0, M3/3. Dental formulae are of the form Inn, Cnn, Pnn, Mnn. I is for incisor; C, canine; P, premolar; and M, molar. The first n is the number of upper teeth on one side, and the second n is the number of lower teeth on the same side. To calculate the total number of teeth, add all of the n values together and double the resulting sum.

Since rodents have no canines or premolars, this leaves a big space (diastema) behind the incisors and ahead of the molars. Rats have a fold of cheek tissue that can pass through the diastema and come together, closing off access to the throat. This allows rats to gnaw through coarse materials without swallowing any debris. The beaver is also a rodent that chews through tremendous amounts of wood both above and below the water's surface. Closure of the throat opening protects the beaver from swallowing wood chips or water.

The rat can move the lower jaw forward or backward so that the lower incisors may be either ahead or behind the upper incisors while gnawing. The front of the incisors is made of hard enamel while the rear is made of softer dentine. While gnawing in these two jaw positions, this arrangement enables the enamel to rub against dentine. Wearing down dentine at the back of the teeth and not the enamel in front sharpens both upper and lower incisors so they act like sharp chisels. The same sharpening technique is used by the beaver.

Rabbits were classified in the order Rodentia (rodents) until the 1920s when they were reclassified in the order Lagomorpha. Lagomorphs have the dental formula I2/1, C0/0, P3/2, and M3/3. Thus, lagomorphs have four upper incisors and two lower incisors. Rodents have two upper and two lower incisors. This distinguishes rabbits from rodents. Rabbits have a peculiarity in that behind the two central upper incisors, there are two small "peg teeth." Lagomorphs differ from rodents in three other respects: rabbits are completely herbivorous; they have no bone (baculum) in the penis; and the scrotum of the male rabbit is ahead of the penis.

Elephants have four molars in their mouths at any one time, one in each of the positions, upper and lower, right and left. A second set of four molars may be partially erupted from the gum. These molars are formed from vertical plates. As the current set of four molars wears down and falls out, a new set of four advances from the rear of the mouth. This process continues until the sixth, and last set of four molars, is in place. Each time a new set comes forward, the molars are somewhat larger to provide for a larger elephant.

Elephants never really stop growing. An elephant's molars weigh 11 lb. (5 kg) and are the size of a large brick. It is common for mammals to have baby teeth (deciduous teeth) which are replaced by permanent teeth, but having five replacements as with elephant molars is unique among mammals.

When the last set of molars is in place, the elephant changes its diet from grass and tree branches in favor of softer aquatic plants. In the wild, elephants are known to eat branches of the acacia tree covered with two-inch thorns[9]. Giraffes and camels also eat these thorny branches. If an elephant cannot reach the highest tree branches, he may simply push the tree over. The maximum life span of elephants is limited by how long their molars last. When its last set of molars wears out, the elephant starves to death[10]. In Kenya, elephants live sixty to sixty-five years, while in Tanzania it is seventy to seventy-five years. In Tanzania, there is more grass available, whereas in Kenya, elephants must browse on leaves and tree branches in addition to grazing on grass.

Mother elephant playing with her baby

Grass is more forgiving of dental wear than tree branches are.

This behavior of old elephants led to an accumulation of elephant bones in the same limited aquatic area. This gave rise to the mythical "elephant graveyard," a place where old elephants went to die. While we are on the subject of unique features of elephants, let me mention that they are the only animals whose knees bend forward.

Though the kangaroo is a marsupial and not a close relative of the elephant, the kangaroo has similar dental challenges[11]. Kangaroos also feed on the tuff grass. Kangaroos have sixteen molars, eight in each of the rear quarters of the mouth. Unlike the elephant, all sixteen molars are present in the mouth at the same time. Just as with elephants and mastodons, molars

---

9   In Berlin, Germany, there is a program encouraging zoo patrons to donate Christmas trees to be fed to elephants. The elephants seem to enjoy eating the trees, and it helps to clean their teeth.

10   The same fate awaited the extinct mastodons, which looked like hairy elephants. They too had but four molars in their mouth at one time. Their molars were made up of vertical plates like those in elephants. When the mastodon's sixth set wore out, the mastodon starved.

11   Another marsupial, the Virginia opossum has fifty teeth, the most teeth of any land animal. This species of opossum is the only North American marsupial.

advance from the rear as the foremost molars wear out and are discarded. The kangaroo starves once it has used up its rearmost set of four molars.

Cows, goats, sheep, giraffes, and ruminants, generally, have lower incisors but no upper incisors. It comes as a complete surprise to most people that cows have no upper incisors. In place of upper incisors, they possess a hard palate. Cows are not without their dental problems. Grass with its silica coating challenges the dentition of all herbivores. A herd of dairy cattle should be inspected every year. Any cow with molars worn down to the gum line should be culled, that is, made to contribute to the gross national product (hamburger).

Horses have both upper and lower incisors and are able to clip grass down to ground level. Horse's teeth grow throughout the life of the horse. This has inspired the phrase "getting long in the tooth." This is usually used in reference to people, though human teeth do not continue to grow throughout life.

The teeth of horses also wear down with chewing. The horse grinds its teeth laterally when it chews. The lower jaw of a horse is narrower than the upper. This leads to incomplete lateral motion so that the medial side of the lower molars fails to wear down. This causes a sharp wall to build up medially along the lower molars. At the same time, a sharp wall builds up along the lateral side of the upper molars. As the teeth continue to grow and wear down, the sharp walls get higher. This interferes with chewing. Grain may be seen falling from the horse's mouth. The sharp edges can cut into the horse's cheeks and tongue, and he may bolt his food without adequate chewing. This can lead to digestive problems. The horse may resist eating and become malnourished. Horses should be seen by a veterinarian approximately annually so that he can file down these sharp ridges. This process is called floating, which employs a tool called a float (rhymes with boat). Floating is not painful since the horse's teeth do not have nerves, but the process may be stressful and may require sedation.

Sometimes the upper molars may be slightly forward than the lower. This leads to an analogous formation of hooks at the forward and rearward ends of the molars. A downward "hook" develops in the foremost upper molar and an upward "hook" on the rearmost lower molar. These hooks need to be clipped under sedation by the vet.

On the upper corner incisor in the horse, there is a groove called Galvayne's groove. The length and condition of this groove give a good indication of the horse's age. The groove begins to disappear at age twenty and is gone by age thirty.

Here's another dental concern for horse breeders; between the horse's incisors and premolars is a relatively free space which horse people call the bars or interdental space. Each of these four spaces, top and bottom, right and left, normally has just one tooth, the canine. Canines normally occur only in males and erupt when the horse is four or five years of age. Often in males, there is a small "wolf" tooth in the interdental space, almost always in the upper arcade, that is, the upper set of teeth. Wolf teeth rarely occur in mares. This tooth is vestigial, that is, inherited from some extinct ancestor, probably not a wolf. I don't know why they are called wolf teeth. The bit sits in the interdental space and is likely to cause pain when it hits a wolf tooth. The solution? Another vet bill. The tooth probably should be removed.

Not all of the food taken in by an animal ends up as carbon dioxide and water. Solid waste leaves from the intestine by the anus in mammals or cloaca in birds, reptiles, and fish. Horse feces tend to be in round balls, very unlike a cow paddy. The horse feces contain a lot of rough undigested material and are relatively dry. The horse is a hindgut digester with a relatively rapid transit time of the digesta. This leads to incomplete digestion with relatively dry feces. Cow feces are much more liquid and produce a flat semiliquid cow paddy. This just may be due to the 30-35 gallons of saliva swallowed by the cow in one day, which enables the cow to provide good mixing in its large fermentation vat. Both horses and elephants, also hindgut digesters, produce almost spherical droppings which contain much undigested plant stems.

Rabbits are hindgut digesters and produce two types of fecal pellets. While actively feeding, the pellets are hard and dry. While the rabbit is actively feeding, food passes quickly through the system, and bacteria do not have time to make certain vitamins. While resting at night, rabbits produce softer pellets which are generated by bacterial fermentation in a blind sac of the large intestine, the cecum. The pellets produced in the cecum are called cecotropes. These pellets are softer and are encased in a proteinaceous coating. In the morning, the soft cecotropes are consumed without chewing as soon as they are passed. The soft pellets are partially made up of the bacteria which carry out fermentation. This habit provides the rabbit with essential vitamins which were not produced as food made a quick first pass through the rabbit's system. At night, food spends more time in the cecum so that bacteria have an opportunity to produce the needed vitamins. These types of feces are the rabbit's only source of vitamin B12.

The young of several herbivores eat the feces of their mother in order to colonize their gut with bacteria needed to accomplish digestion. Baby pandas, koalas, hippos, and elephants all do this. As the koala is marsupial, not a bear (as it is sometimes called), the newborn is about the size of a small jelly bean and weighs less than a gram. It does not eat its mother's feces until it is about four months old and is ready to be weaned. Marsupials, by definition, produce young at an extremely early stage of development. Most, but not all marsupials, have pouches in which the young spend the early part of their development.

Nitrogen is a component of amino acids, the building blocks of all proteins. In herbivores, these acids are supplied by bacterial fermentation. The bacteria themselves grow and multiply inside the herbivore's body and constitute a significant part of an herbivore's diet. The nitrogen leads to toxic wastes which must be excreted. Nitrogenous wastes are excreted in the urine or feces, usually in one of three chemical forms: ammonia, urea, or uric acid. Animals excreting these wastes are described as ammonotelic, ureotelic, or uricotelic[12] respectively.

Fish and crocodilians (includes alligators) excrete nitrogen as the toxic compound, ammonia[13]. In order to avoid toxic reaction to ammonia, the compound must be greatly diluted with water in ammonotelic animals. Ammonia excretion carries with it loss of copious amounts of water, but aquatic animals have an abundant supply of water. Excretion of ammonia occurs through gills, skin, or kidneys.

Mammals, including man, and adult amphibians (frogs, newts, and salamanders) excrete nitrogenous wastes as urea. In frogs, urea is mixed with intestinal wastes and exits the body via the cloaca. Urea excretion is more sparing of water loss than is excretion as ammonia. Ureotelic sharks retain urea in their blood so that their blood has a lower concentration of water than is present in seawater. This promotes the movement of water from the sea into the bloodstream by the process of osmosis. This absorption of water keeps the shark hydrated. The shark can excrete any excess water as dilute urine.

Reptiles, birds, and insects excrete nitrogenous wastes as uric acid. Uric acid is produced by enzymes in the liver, acting to detoxify ammonia. Uric acid is also a metabolic product of the DNA present in a meal containing meat. Turtles are both ureotelic and uricotelic, depending on circumstances such as the supply of water. The uric acid excreted by turtles is found in solid crystals in their cloacae, precipitated from urine. Sea turtles maintain proper levels of sodium chloride (table salt) by an organ which pumps out the salt in tears. Seabirds, considered by most scientists to be reptiles like turtles, also shed salt tears. Birds do not urinate to rid themselves of uric acid. It is passed by birds and turtles in their feces. You may have been asked, "What's that white stuff on top of chicken dung?" The answer is not "more chicken dung." Now you know; it's uric acid. This saves water, not always in plentiful supply.

---

[12] The -telic part of these words refers to "distant" (final) as in telephone or telescope.

[13] In keeping fish in aquaria, it is good advice to "break in" the aquarium with inexpensive fish. Ammonia excretion can build up in the aquarium and poison fish in a new tank. As the tank ages, beneficial bacteria convert ammonia to nitrite ion and later to nitrate ion. In going from ammonia to nitrite to nitrate, toxicity decreases. Once you reach the nitrate stage, it's time to buy some expensive fish.

The dalmatian dog also excretes uric acid; other dog breeds excrete urea[14]. Finally, scorpions and spiders excrete nitrogen as guanine, which contains five nitrogen atoms per molecule, as opposed to four in uric acid, and thus saves even more water. Scorpions and spiders often live in deserts.

Camels are able to recycle their urea production into making protein. Instead of excreting urea in their urine, camels are able to direct urea into their stomach where bacteria hydrolyze it to carbon dioxide and ammonia. The bacteria then use the ammonia to make protein. Camels are able to subsist on a virtually protein-free diet of low-grade hay and dates.

---

[14] Dalmatians should not be fed organ meats as these foods produce uric acid, which can deposit as stones in the bladder. About 12% of dalmatians are born deaf. Organ meats should be restricted from the diets of human sufferers of gout, caused by the deposition of uric acid crystals in joints. Uric acid has low solubility in water.

Chapter 4

Predator-Prey Relationships

Exothermic predators such as crocodiles, rattlesnakes, or lizards tend to hunt by using a "sit and wait" strategy. They do not have the stamina to go on long hunts for their prey, so they let their prey come to them. Exothermic predators are unable to give chase for long distances, but can accelerate quickly and run fast for short distances. This effort is anaerobic (not using oxygen), and lactic acid builds up and leads to muscular fatigue. The exothermic crocodile can move with lightning speed to catch a wildebeest drinking at the water's edge. The crocodile drags the wildebeest under the water and drowns the animal. Other crocodiles may assist in dismembering the prey by holding the body steady while the attacking crocodile rotates its body until a limb is twisted free from the victim. The crocodile then swallows the limb whole. Crocodilians have higher acidity in their stomach than any other animals do and are able to digest bones. The other crocodiles behave in the same manner. The crocodiles share in the bounty without dispute. When a crocodile is exhausted from such efforts, it is unable to move and must rest and take in oxygen which is needed to rid itself of accumulated lactic acid. Until the crocodile has recovered, he is completely harmless.

When forests receded some twenty-five million years ago in Africa and were replaced by grasslands, grazing animals took advantage of the ecological niche provided. They developed teeth and digestive systems appropriate for eating grass. This meant having large stomachs and intestines to handle the large quantities of grass necessary for their nutrition. Grass is a poor source of nutrition, but you have to take what's available. These herbivores (plant eaters) developed long legs so they could cover enough territory to harvest a lot of grass.

This situation provided another ecological niche. With all of these large animals converting low-quality food like grass into high-quality protein, why not eat them? Of course, that's what happened. Along came lions, leopards, cheetahs, hyenas, and wild dogs. Not to let anything go to waste, vultures were on hand to clean up the leftovers. The grazing animals, for example, wildebeests, soon found that they fared better at staying off some lion's menu if they grouped in herds. They could place their young in the middle of the herd for protection. The presence of so many animals in one place gave each individual a better statistical chance of survival. Were wildebeests good statisticians? No, they reasoned inductively. They simply observed that staying together worked out better than being a big helpless animal, alone in the grass, with almost nowhere to hide. The long legs of the herbivores came in handy for fleeing from predators.

Giraffes showed up with long legs and necks, enabling them to eat the leaves of the occasional trees on the savanna. Wildebeests made friends with giraffes. They loved having a fifteen-foot lookout to warn them of approaching lions or hyenas.

A wrinkle in these relationships comes about in theft of kills by competitors. A group of hyenas will very often harass a lion who has brought down prey until the hyenas drive the lion away, surrendering its kill. Lions return the favor by stealing hyena's kills. Winners and losers in these contests are usually determined by relative numbers of contestants. Among cats, only the lion tends to hunt in packs like the wolf does. Female lions do the hunting. Males are better at eating. The males are also better at defending the kill against hyenas. The leopard has his own way of defending his kill. The leopard, a very good tree climber, hoists the carcass up a tree and lodges it in a fork in the branches.

The cheetah is the fastest land animal on the planet. Cheetahs can run 70 mi./hr. (112 km/hr.). Cheetahs have nonretractable claws like those of a dog. This gives them better traction while running. Lions have to get close before they attack in order to be successful. Most of the time they fail in their attempts and sometimes they end up starving to death. Cheetahs can attack from a little farther away. The prey animals know this and respond appropriately. They are not worried about lions until the cats get too close. Don't let the cheetah get so close!

Prey animals like gazelles sometimes leap straight up in the air when threatened by lions. This action is called stouting. This looks like a counterproductive action, which should tire the gazelle to no useful purpose. Stouting sends the following message to the lion: "Don't bother wearing yourself out chasing me because, as you can see, I am very physically fit. You will end up tired and hungry because you can't catch me." The message is sincere because stouting is, in fact, tiring. Lions normally don't chase stouting prey.

Chapter 5

Animal Senses

Animals interact with their environment through their senses. The senses of a particular animal are evolved to meet the needs of that creature. We may admire those animals having particularly sharp senses, but what really matters to the animal is how well its sensory system equips it to deal with life's challenges.

Many animals possess a special organ near the roof of their mouth called the vomeronasal organ (VNO), which allows them to detect pheromones and other materials in the air. Some widely divergent animals having the VNO include snakes, mice, rats, elephants, giraffes, goats, pigs, and dogs. Snakes especially are in constant touch with their environment in this way. When they flick their tongues, they are sampling the air. Snakes transfer the air sample to the roof of their mouth in order to contact the VNO. The VNO is also called the Jacobson's organ, after Ludvig Jacobson, its Danish discoverer. The VNO is often used to detect pheromones which are chemical signals that a female is receptive to mating. The VNO may also detect odors not associated with mating. Painted turtles use their VNO to provide a sense of smell under water. Cats and hoofed animals (ungulates) exhibit a distinctive facial gesture called the flehmen response to engage the VNO. This may be observed in the house cat. When the cat detects a smell of interest, it holds its head up, stops breathing momentarily, and curls its upper lip upward. Elephants gather scents onto a fingerlike appendage at the tip of their trunks and touch it to the VNO in the roof of their mouth. The African elephant has two fingerlike appendages at the tip of its trunk; the Indian elephant has just one. The African elephant is able to pick up small objects with these fingers.

The male giraffe stimulates the external genitalia of the female, inducing her to urinate. He catches some of the urine on his 18 in (46 cm) tongue and transfers it to his VNO. In this way, the male giraffe is able to detect the female's sex pheromones if the female is in estrus. If she is receptive, he may follow her for a couple of hours, after which she may permit copulation lasting about ten seconds. Fifteen months later, she delivers offspring while standing up. The baby giraffe falls 6 ft. (2 m) to the ground, instantly severing the umbilical cord.

Male giraffe in African grassland

A few of my favorite things—whiskers on kittens, rats, and fur seals.

Whiskers on kittens. A house cat's whiskers can be used to gauge whether it can fit through an opening, but the cat is much more dependent on its whiskers than just for that. A blind cat can get along pretty well by virtue of its whiskers. Cats have two rows of whiskers on their muzzles with twenty-four whiskers in all. Besides these, the cat has whiskers above its eyes, on the back of its legs, and also on top of its head. These whiskers are acutely sensitive to air currents and vibrations in the air.

The cat has very sensitive ears which pick up sounds due to vibrations in air. A cat can hear the high-pitched squeak of a mouse outside the range of the human ear. It can also sense these sounds with its whiskers. Cats can tell from air currents the size and shape of obstacles such as furniture in their path. This is useful especially at night. Though cat's eyes are specially adapted for night vision, they can't see at all in total darkness. Cats are farsighted and may not see well the mouse already in their grasp. The cat's whiskers can tell it if it has a good hold on the mouse.

A cat's canine teeth are capable of inflicting a bite precisely designed to sever a prey's spinal column. The cat's whiskers tell it where to bite. A cat's whiskers can be an annoyance. If the whiskers touch the sides of a bowl from which it is eating, its brain receives irritating messages, making it difficult for the cat to continue eating.

Cat whiskers fall out from time to time, but grow back after a while. Each whisker is connected by a nerve directly to a specific site in its brain called a barrel. The site is called a barrel because of its shape. Each whisker has its own barrel. If whiskers are removed, the barrels shrink. If whiskers grow back, the barrels recover at least partially. Never trim a cat's whiskers. This is traumatic for the cat, bringing on a state of depression from which the cat may never completely recover. Cats continually move their whiskers in order to gather sensations much as a blind person might use a cane.

Whiskers on rats. Rats are possibly even more dependent on their whiskers than are cats. Rats are mainly nocturnal animals and have weak eyesight. They tend to seek out dark places such as the interior of walls and sewers. Like cats, they can find their way by sensing vibrations and currents in the air. Just like in cats, the shorter hairs are tuned to higher-pitched vibrations and the longer hairs to lower-pitched vibrations. They move their whiskers toward an object they wish to investigate. They drag their whiskers across surfaces over which they are traveling and sense the texture in detail. They are able to sense very small bumps or scratches in a surface.

Whiskers on fur seals. The northern fur seal, which as you are probably aware was subject to decimation for its valuable fur, is strongly dependent on its whiskers. It makes deep dives up to 820 ft. (250 m) to chase its prey, mainly fish, into dark, murky waters. There is no light at these depths. The northern fur seal has whiskers, which enable it to sense the weakest currents in the water produced by evasive mackerel. The nerve signals generated from the whiskers enable the seal to locate its prey in total darkness.

Elephants are able to hear sound frequencies lower than are detectable by the human hearing. Elephants are able to communicate over many miles of separation. Low-frequency sounds travel long distances much more efficiently. Shorter-wavelength sounds are scattered by objects on the landscape. Whales make use of this phenomenon. Water conducts sound more efficiently than air does, and whales can communicate over distances of thousands of miles. In shallow water there is so much noise in today's oceans that communication between whales is severely impacted. Whales are probably missing many mating opportunities. Bats, porpoises, and shrews can send out high-frequency sounds and detect objects by sensing the reflections of this ultrasound. Hippopotamuses produce clicks and whistles similar to those of porpoises and are now believed to be most closely related to porpoises and whales.

Chapter 6

Sleep Habits of Various Animals

First, a few words about the nature of sleep. Sleep and the necessity for sleep are not well understood. What we do know about sleep is that it is not a simple period of time during which everything shuts down for the night. There is a circadian (daily) rhythm which influences the daily lives of all living things, plant and animal. Along with sleep cycles, there are regular body temperature changes. Our temperatures are highest at about 6:00 PM and lowest at about 5:00 AM. We tend to go to sleep when our temperature starts to fall from the maximum and tend to awaken as the temperature begins to rise after its minimum. This fluctuation amounts to about 0.5°C (0.9°F). You may have noticed that when we run a fever, our temperature generally rises at bedtime and is lowest first thing in the morning. When first we fall asleep, we pass into a stage of deep, dreamless sleep. During this period, the heart rate is slow, and breathing is slow and regular. This is followed by a stage called REM sleep. REM is an acronym for rapid eye movement. In REM sleep, both breathing and heart rates are irregular as though the sleeper were emotionally upset. In contrast to an emotional state, the muscles in REM sleep are relaxed. This apparent contradiction of emotional appearance along with muscular relaxation prompted early investigators to call this stage of sleep "paradoxical sleep". It is during this stage of sleep that intense dreaming occurs. Though this is an active stage of sleep, it is the stage most conducive to restful refreshment of body and mind. In order to feel well rested, it is necessary to achieve REM sleep. There are several cycles of deep and REM sleep each night. The most time spent in REM sleep occurs during the latter half of the night. Persons deprived of

REM sleep tend to increase the percentage of time spent in REM sleep to make up the deficit[15].

The sleep habits of animals are often a reflection of the requirements of life's activities. Some shrews hardly sleep at all. They are busy day and night gathering food to satisfy their high rate of metabolism and to keep their tiny bodies warm, as they are endothermic. For shrews to sleep through the night would mean death by starvation. Like hummingbirds, shrews are always on the edge of starvation, but while hummingbirds estivate at night, shrews eat all night and never sleep.

Elephants and giraffes don't get much sleep either. They have to forage for low-calorie food during most of a twenty-four-hour cycle. The elephant sleeps about four hours a day and the giraffe only about two hours. Both elephants and giraffes recline on their sides when sleeping. In order to achieve REM sleep, elephants and giraffes must recline because during REM sleep, muscles are relaxed and cannot support the animal's weight. In zoos, elephants are light sleepers, awaking at slight disturbances and failing to return to sleep quickly. Horses, on the other hand, can and do remain standing in their stalls while asleep. The horse has a special anatomic feature which allows the legs to lock in place to provide support while asleep. A full-grown horse is more comfortable standing than lying down because when reclining, his weight presses on his internal organs.

Koalas, which are warm-blooded marsupials, spend about fifteen hours a day in slumber in order to conserve energy. These arboreal creatures eat practically nothing but the leaves of the eucalyptus tree, a diet providing scant nutrition which would be toxic to most animals. Their poor diet leaves them with little energy.

The echidna, which has a low body temperature, does not exhibit REM sleep. Another mammal with low body temperature, the armadillo, sleeps over seventeen hours a day of which about 18% is spent in REM sleep.

The hippopotamus is capable of sleeping in the river where it spends the day after grazing on land during the night. The hippo can sleep submerged, but comes up every fifteen minutes or so to breathe. Its nostrils and eyes as well are situated high on its head so that it can breathe submerged except for

---

[15]   Thomas Edison felt that sleep was a waste of time and tried to minimize the amount of time he spent sleeping. He used to take sleep in a chair, holding a metal ball in his hand. As REM sleep approached, his muscles would relax, and he would drop the metal ball into a metal bucket, and the clatter would awake him. Ideas about REM sleep did not develop until well after Edison's death.

its nose and eyes, which are closed in sleep. The hippopotamus can also close its nostrils when submerged.

Northern fur seals have a very interesting method of repose when they are far out at sea. The fur seals remain afloat while sleeping by using one of their front flippers to gently swim in a small circle. They use first one flipper and then, after perhaps twenty minutes, they switch to the other. Researchers believe that the northern fur seal allows one hemisphere of its brain to go to sleep

Northern fur seal on rocky shore

while the other hemisphere engages the flipper to keep his nose out of water. The left hemisphere controls the right flipper, and the right hemisphere controls the left flipper. The two sides of the brain take turns being awake or asleep.

Chapter 7

Tampering with Nature

Unfortunately, man has often tampered with nature either accidentally or deliberately. There are many examples of human intervention leading to the introduction of animals into environments far beyond the animal's normal range. One prominent example which has led to some of the greatest suffering and death that has ever afflicted mankind was the spread of rats. Rats have also done and continue to do tremendous harm to many wild animal species. Early explorers from Spain, Portugal, Britain, and Holland transported rats on probably every ship that sailed. This spread rats to every place on earth except the Polar Regions.

The two most common species of rat are the black rat and the brown or Norway rat. The black rat gained everlasting notoriety by carrying bubonic plague which decimated Europeans during the Renaissance. Rats harbored fleas (*Xenopsylla cheopis*) infected with *Yersinia pestis* which is the putative agent for the bubonic plague-black death. The adjective *black* refers to buboes which showed up as black lumps in the lymphatic tissues. Repeated outbreaks killed perhaps a third of the population of Europe. Occasionally, *Y. pestis* is still found in ground squirrels at the foot of the Mount Sandia[16] in Albuquerque, New Mexico.

Black rats are sometimes referred to as roof rats as they frequently seek higher floors in buildings. Black rats (*Rattus rattus*) have often been displaced by competition from the somewhat larger and more aggressive Norway or brown rats (*Rattus norvegicus*). In some places, the black rat could be considered locally threatened. The name Norway rat was coined by a British

---

[16] Sandía is Spanish for "watermelon." The mountain has the color and shape of a lengthwise-sliced watermelon as viewed from the west at sunrise.

naturalist who believed that the brown rat came from Norway in a shipment of lumber. Actually, the Norway rat was present in Britain before it spread to Norway[17]. The Norway rat is believed to have originated in Asia.

The Norway rat is the common house rat familiar to everyone. "Build a better mouse trap and the world will beat a path to your door." This quotation from Ralph Waldo Emerson could equally apply to the familiar Norway rat. This calls to mind the titanic struggles that have been going on for centuries to rid the planet of rats. It also recalls the strategies the rat has used to counter human efforts without expending any thought, but just doing what comes naturally to a rat. Rats are notoriously xenophobic. They distrust anything new in their environment such as a rattrap that wasn't there yesterday. Pest-control personnel sometimes leave unbaited traps lying around to familiarize the rat with them before setting the traps.

It may come as a surprise that rats are distrusting of unfamiliar foods, knowing that rats will eat almost anything. They first like to study a new food and perhaps take a little bite. There is good reason for this fear. Rats, like horses, cannot vomit and so, like horses, are especially subject to being poisoned. If a young rat sees Mom eat something, it knows the food is safe. If a rat sees another rat eat something and become sick or die, the rat will avoid this substance. In the pest-control business, this is known as bait shyness. To counter bait shyness, pest-control businesses hire chemists to develop substances which will kill the rat slowly. In this way, his fellow rats won't be able to connect his demise with the food he ate several days ago. The first significant rat poison came about in an interesting way.

> Maud Muller, on a summer's day,
> Raked the meadows sweet with hay—

Opening lines of a poem by John Greenleaf Whittier, an American poet (1807-1892). The fragrance of sweet clover is due to the organic compound coumarin. Dairy farmers are vigilant in putting up hay to see that the mown hay is quite dry before being stored in the haymow. If the hay is stored green or damp bacterial action converts the coumarin to dicumarol, which causes internal bleeding in cows that eat this hay. A closely related compound, warfarin (also pharmaceutically known as Coumadin), causes internal bleeding in rats, killing them slowly. Warfarin was the first successful rat poison.

Warfarin is used in medicine to prevent the formation of blood clots, which can travel in the circulation and lodge in the brain or heart, causing

---

[17] Chares Dickens in 1888 realized this mistake and wrote of it in his weekly journal *All the Year Round*.

a stroke or myocardial infarction. Doctors, perhaps wisely, don't always tell their heart patients that they are being treated with rat poison. The dosage must be carefully adjusted in patients so as to be effective without undue risk of bleeding from a minor wound or a small stomach ulcer. To kill rats, the warfarin is mixed with food to produce a "bait poison."

Rats have not studied Darwinism, but they don't need to. Darwinism comes naturally to a rat. By the process of natural selection, rats, somewhat tolerant to warfarin, have produced offspring a bit more tolerant, until warfarin is now of little use in combating rat infestations. We are used to hearing about AIDS viruses developing drug tolerances so that multidrug therapy (MDT) is the only way to treat AIDS. Even with MDT, AIDS is not yet curable. The same is true for the treatment of the bacterial infection tuberculosis, which is curable with MDT.

Chemists have come up with stronger anticoagulants (brodifacoum) to combat rat infestations and sometimes are using multipoison treatment to kill rats. There is a downside to brodifacoum. Though more poisonous to rats, it is also much more poisonous to humans than warfarin. Keep out of reach of children! It is very difficult for an animal or even a virus or bacterium to develop resistance to several poisons at once. Viruses and bacteria have an advantage not shared by animals in the development of drug tolerance. They produce new generations very quickly, and with each new generation comes an opportunity to counter the poison. Rats don't reproduce as quickly as bacteria do, but they are fast reproducers, to be sure. The gestation period for the brown rat is twenty-one days, and brown rats typically have litters of seven.

Campbell Island, New Zealand, was once the most heavily rat-infested place on Earth. After its discovery in 1810, it was used as a base for sealing and whaling. Campbell Island is almost halfway to the Antarctic coast from New Zealand. The weather is fiercely cold and windy. On Campbell Island, there were three attempts at sheepherding from 1896 to 1931, when the last leaseholder and his employees were rescued from starvation. The remaining sheep were killed by order of the New Zealand government to save them from death by slow starvation. The last one was shot in 1992.

Norway rats had killed all of the native land birds and smaller seabirds. Most of these had not yet been classified by zoologists. Often, bird species on small isolated islands have no natural predators in the absence of rats. They do not guard their nests with care, and many have lost the power of flight or are in the process of losing flight. Without predators, the birds don't need to fly away. Rats feast on eggs and chicks as well as adult birds. Campbell Island is not the only example of extinctions due to rat infestations. It is a common occurrence. Authorities believe that the dodo, which lived on Mauritius Island, was exterminated by black rats. Of the thousand most endangered species, according to the Nature Conservancy, 42% are at risk due to invasive species, usually rats.

The New Zealand government was eager to restore wildlife to Campbell Island. In June 2001, Pete McClelland headed a team of men charged with ridding the island of the two hundred thousand rats living there. These rats were descendents of rats that had come ashore from every sailing vessel that had ever landed on the island. Wherever ships went, rats went with them. The rats were eliminated by poisoning with tons of brodifacoum over a period of two years. Pete and his team started in the winter when food was in scarce supply and rats would be hungry. A final cleanup operation involved the services of a pack of specially trained dogs. In May of 2003, the government of New Zealand officially declared Campbell Island free of rats.

The flora of the island, in particular the grass that grew slowly in this cold climate, came back nicely. The fauna also recovered remarkably. Even a species of duck, the Campbell Island teal, which had been considered extinct, has, through a captive breeding program, been restored. This teal has been upgraded from "critically endangered" to "threatened."

Another example of man's tampering with nature involves a strange case of reverence for an animal which led to the near extinction of another animal. The population of vultures in southern Asia is falling to the level of critical endangerment. During the last score years, the number of Oriental White-backed vultures has decreased from tens of millions to as few as eleven thousand. Other vulture species have also decreased to dangerously low levels. The cause of this sudden decrease was mysterious until finally it was learned that vultures were dying due to kidney failure from ingestion of diclofenac, a nonsteroidal anti-inflammatory drug (NSAID).

This drug is used for treating human arthritis pain. Diclofenac is given in India and Nepal to aging cattle to alleviate the suffering of arthritis. In these countries, cattle are considered sacred. They are raised as farm animals for their dairy products, to pull carts, and to till the soil. Cattle not owned by anyone are also free to roam the streets and countryside. India has 30% of the world's cattle. They are not harvested for meat and are even sometimes kept in "bovine hospices" until they die of natural causes. Vultures formerly cleared the countryside of dead cattle.

A vulture-safe NSAID, meloxicam, can be used in place of diclofenac, but it costs two to five times the price of the latter. Furthermore, the human pharmaceutical diclofenac is less costly than the now-outlawed veterinary drug. Manufacture of veterinary diclofenac became illegal in 1996, but there exist stores of this long-shelf-life drug so that it is still sometimes given to cattle. India is ill equipped to incinerate or remove in other ways all of these animal carcasses. The result is a danger to public health. There have been two outbreaks of anthrax attributed to dead cattle.

There is a small Parsi ethnic minority whose religious tradition forbids them to bury their dead. Their custom was to lay the dead in "towers of silence," where vultures could consume the bodies. With a paucity of vultures, they have had to use large solar reflectors to speed bacterial decomposition.

The subject of man's interference with animal welfare could easily support a thousand-page exposition. We must be content with this little scratch on its surface.

Chapter 8

Styles of Reproduction

What causes species to change from one generation to the next until new species are produced? A few forces at work were well understood by Charles Darwin:

(a) There must be differences between individuals in a population (biological variation). If this were not so, there would be no basis for selection.

(b) Survival to breeding age and capability must not be assured. Usually, many more offspring are produced than can hope to survive. This ensures that there will be a competition which will be won by the fittest.

(c) There must be a mechanism for selecting individuals for desirable traits.

Requirement (c) leads to many routes of satisfaction. An obvious selection process occurs in choosing a mate. The female rejects one suitor and chooses another. Males fight for dominance, and the dominant male breeds, while the loser does not. Predators eat the slow animal, and the fast one escapes. Man encroaches on habitat. Some animals adapt very well to man's presence, for example, rats, cockroaches, and raccoons. Others do not. The climate changes. Some adapt by seasonal migration; others do not. The litany goes on and on.

Some animals such as frogs and fish lay hundreds of eggs which are fertilized by the male immediately after being deposited. This strategy of reproduction relies on survival of sufficient individuals based on sheer numbers. In some cases, the young are cared for, and in others, no effort is invested by the parents beyond egg laying and fertilization.

Reptiles such as snakes, crocodilians (alligators and crocodiles), and birds lay eggs. In some schemes of classification, birds are quite reasonably considered to be reptiles. Birds share many traits common to dinosaurs. It has been established that some bird DNA is also close to that of dinosaurs. Some of this vestigial dinosaur DNA in birds has been activated in research laboratories.

In all of the above-mentioned animals, the eggs develop inside the female's body and are subsequently laid and incubated by body heat (in birds) or by environmental warmth for snakes, crocodilians, or turtles. In the case of crocodilians, warmth of decaying vegetation in the nest serves to incubate eggs. While eggs are incubating, they have to be guarded to prevent predation. After hatching, there is usually more effort expended in protecting offspring from predators and sometimes greater effort in feeding the young. Unlike in frogs or fish, there is a relatively great investment in each egg the female produces. Because sperm are usually produced in the millions, there is little effort invested in each sperm.

In most mammalian species—such as cattle, dogs, cats, beavers, or humans—only one or a few eggs are fertilized during a breeding season. Again, there is little investment in a given sperm among the millions released. In typical mammals, the fetus requires an extended period of time to develop in the womb. The female must provide nourishment before birth, and after birth, the mother must produce milk and feed the young by nursing. She must invest much time and effort in this process. The male may invest relatively less in the reproductive process. We must realize, however, that the male may invest considerable effort in fighting for a mate. These contests may also involve risk of injury. Often during rut, the male goes without eating.

It is important for the female to produce the healthiest offspring possible. Part of her success in producing viable offspring is to choose the strongest, healthiest mate possible. For this reason, it is normally the female who chooses which male will be her partner in procreation. She tends to be choosy in selecting a mate; after all she is going to make a tremendous investment in the reproductive process. On the other hand, the male is capable of inseminating many females. He is not choosy for his investment in reproduction is much less than the female. He may try to mate with as many females as possible. If he is a "good catch"—that is, big, strong, and handsome—he may have a harem of several females and he will guard and protect his females, repelling other males with all his strength. The female's mating and parenting behavior seems to be geared to producing the healthiest offspring. This gives her the best chance of passing on her genes. The same can be said of the male's behavior. Both male and female seem to be trying to preserve the species, but this may not be the case.

Consider the behavior of male lions. A pride of lions consists of a matriarch, her sisters, older offspring, and their cubs. There may be one or perhaps two or three male lions associated with the pride. Sometimes these males may be challenged by males from outside the pride, and they may be displaced. If new males take control of the pride, bloodshed follows. The new males bite the necks of the cubs, breaking their spinal columns. The new males are not concerned with protecting cubs not their own. It is as though they want to pass on their genes, not those of another lion. Do you think the lion knows that he has genes? Wouldn't sparing the cubs help survival of the lion species? This apparent attitude is common in the animal kingdom. Bears do the same. Animals don't have to understand anything about evolution. Whatever works is what survives. The animal does not have to understand how it works. Perhaps this behavior is due to the aspect of competition with success for the fittest. The lions who took charge of the pride proved to be the fittest. Perhaps survival of the fittest serves preservation of the species in the long run, even though lion cubs had to die. There are also numerous tricks by which a male tries to prevent another male from breeding with his mate. The most common trick is guarding the female or females in order to repel any competing males. The osprey (fish-eating hawk) mates with the female as many as one hundred times a day, giving other males little chance of access to her.

Some birds have a unique way of making reproduction easier and perhaps more productive. They are brood parasites. They lay eggs in the nests of other species of birds. If their strategy works, the host bird takes on the tasks of incubating the foreign egg and raising the resulting chick. The common (formerly European) cuckoo lays eggs in the nests of many other species, among which is the reed warbler. The cuckoo is larger than the reed warbler. Furthermore, the cuckoo's egg hatches before those of the warbler. The cuckoo chick uses a depressed area between its shoulders to capture a warbler egg or a chick and hoist it out of the nest. Once the cuckoo chick is alone in the nest, the reed warbler parents feed the cuckoo. The parents don't catch on even though the cuckoo grows to be much bigger than his foster parents. The chicks of the African honeyguide bird and the South American striped cuckoo, both brood parasites, have special hooks on their bills with which they peck their host's nestlings, killing them.

There is one case in which brood parasitism appears to be advantageous to both invader and host. There is a South American bird called the oropendola that constructs a pendulous nest 3 ft (0.9 m) long at the ends of small branches. The fragile branches cannot support the weight of animals trying to raid the nests. There is a lightweight predator that can work its mischief on oropendola chicks. Botfly larvae invade the chick's bodies. To defend their brood, the oropendola birds attempt to find nest sites near bee or wasp nests. The bees or wasps use the botfly larvae bodies as food for bee or wasp larvae. Two problems arise in this partnership—the bees/wasps may abandon the site for reasons of their own. The weight of the bee/wasp nest may cause the fragile branch to break. Giant cowbirds act as brood parasites. The cowbird chicks help the oropendola hosts by consuming botfly larvae, thereby reducing mortality by 90%. The cowbird overcomes the difficulty of gaining access to the hanging nest by tearing a hole in the nest and injecting an egg into the hole. The host later repairs the damage.

When cowbirds lay an egg in the nest of a northern cardinal, the cardinal abandons the nest. Cowbird eggs laid in a robin's nest are ejected by the host. Sometimes the would-be host simply builds a false floor over the cowbird egg.

The mating habits of animals seem to explain a great deal about their approach to life. Females seem to be in the driver's seat when it comes to which male will be a daddy. Female birds prefer colorful mates. Their bright colors may lure predators away from the nest. Female frogs prefer deep voices since deep voices mean larger males. Lionesses prefer large males which are better able to protect the pride and to chase away hyenas from a kill the females have just brought down. Because of these preferences on the part of females, it is the big strong, handsome males who get the chance to hand down their genes, and this leads to bigger stronger offspring. (Like father, like son.) Perhaps this is why males tend to be bigger than females.

Most male animals are larger than females are. In the northern fur seal, the difference is so large that biologists at first thought the two sexes were different species. Males are five times the size of females. Males weigh 400-600 lb. (180-270 kg), the females, 65-110 lb. (30-50 kg). Northern fur seal fights between males over mating privileges are particularly vicious. Perhaps this explains the unusual disparity in size of the two sexes.

It would seem advantageous for survival of species that males fight to select the fittest males, but there is little advantage if the fights result in death or serious injury. Fights between nearly all animals, for example, elephants, giraffes, and wolves, while valiantly contested, seldom result in serious injury. The formation of hierarchies (pecking orders) helps in preventing injuries. In wolves, only the dominant male mates with the dominant female. Other pack members do not mate. Normally there is no serious contest once the hierarchy is established. Often hierarchies can be established by harmless sparring among young animals before their first breeding season.

Parenting styles can hasten natural selection. One example involves brown pelicans. Typically, a female pelican will lay two or three eggs. Both parents incubate the eggs and participate in gathering food for the offspring. Pelican chicks are altricial. They are blind and sparsely feathered. The unusual behavior is that the parents only feed one chick, normally the firstborn. The others lose the competition and die. The surviving chick grows to be larger than its parents. At this stage, the parents stop feeding the chick, and it lives off its fat until it learns to gather its own food.

The snowy owl[18] lays her eggs one at a time with one or two days in between. Since the eggs all take the same amount of time to hatch, the chicks are of different sizes. If food is scarce, the eldest chick may eat the youngest. This ensures the survival of the little cannibal.

Eagles sometimes harass female red kangaroos with a joey (young kangaroo) in the pouch until the mother kicks the baby out of the pouch, saving her own energy, but sacrificing the baby. Marsupials have less of an investment in their young than animals that carry their young for long gestational periods.

The Tasmanian devil gives birth to more offspring than she has nipples. Since the Tasmanian devil is a marsupial, the infants are embryonic in development and must have immediate and constant nutrition for survival. Most of the young will lose out in the competition for nutrition and will promptly die. Only the strongest survive.

A similar fate is the lot for most Virginia opossum babies. The Virginia opossum is also a marsupial and gives birth to twenty to thirty or as many as fifty young in a litter, but mother opossum has only thirteen teats. The Virginia opossum is the only marsupial that lives in the United States. Several species of possum live in South America. It is traditional to refer to the North American species as the Virginia opossum, but those in South America are called possums.

Births are said to be multiple when a single pregnancy results in more than one offspring. Presently let us consider the case of two offspring, that is, twins. In the United States, human twins occur in about six of every one thousand births. Twins which result from the fertilization of two eggs with two different sperms are colloquially called fraternal twins. Fraternal twins are related to one another in the same way as brothers and sisters are. They may be of the same sex or of different sexes. A fertilized egg is called a zygote. Fraternal twins are said to be dizygotic (DZ) since they come from two different eggs fertilized by two different sperms. DZ twins have different DNA.

If twins result from the division of a fertilized egg (zygote) into two fetuses, the twins are said to be monozygotic (MZ). Colloquially, MZ twins are referred to as identical twins. Identical twins possess the same DNA and look very similar, but may show differences. For example, they may not be exactly the same size. To some extent, the development of MZ twins is influenced

---

[18] Other facts of interest about snowy owls are that the female is somewhat larger than the male. Also, as is typical of many owls, one ear is placed higher on the head than the other. This improves auditory depth perception so that the snowy owl can pounce on a mouse which is tunneling beneath the snow.

by their environment either in utero or after birth. Environmental influences can result in turning genes on or off. Genes are responsible for production of proteins, including all enzymes. MZ twins may or may not share a common placenta. When the zygote (fertilized egg) cleaves to result in two individuals, development is influenced by the timing of this event. The later the cleavage, the more closely allied will be the two fetuses. If the split occurs within three days of fertilization, each fetus will have its own placenta and its own amniotic sac. If the split occurs from day 4 to day 8, as it does in 60-70% of cases, they share the same placenta but have individual amniotic sacs. If cleavage occurs on day 8 to 13, they will share the same placenta and the same amniotic sac. If cleavage is delayed beyond day 13, the MZ twins will be conjoined, that is, they will be Siamese twins. The original Siamese twins, Chang and Eng Bunker, were born in Siam (present-day Thailand) in 1811.

About 25% of MZ twins are mirror images of each other. Asymmetric features appear on opposite sides of their bodies. As one result, while one twin is right-handed, the other is left-handed. An apparent example of mirror-image MZ twins involved two sets of twins who happened to become world-class professional tennis players. The Bryan brothers—Mike, the older twin, and Bob, the younger—are identical twins who are easily distinguishable. Mike, who is somewhat smaller than his identical twin brother, is right-handed. Bob is left-handed.

The other tennis-playing twins were Tim and Tom Gullikson. Tom is left-handed, and Tim[19] (deceased) was right-handed. A quaint story about the two involved a European player who played in a singles tournament one week against Tim and the following week against Tom. Not knowing they were identical twins, he was heard to say, "Last week he beats me right-handed, and this week he beats me left-handed." This story may seem out of place in a book on animals, but I like the story, and you have to admit that humans are mammals worth considering. That they are animals is undeniable. The only explanation, of which I am aware, for the mirror-image nature of some MZ twins is the following: The twins may be side by side in the womb, and both may face forward. This would place the right arm of one twin adjacent to the left arm of the other. This crowding could influence handedness. It would give a geometric difference which could relate to other asymmetric developments in the twins.

---

[19]   Tim Gullikson died in 1996 of a brain tumor. At the time of his death, Tim was the coach of Pete Sampras, the world's number-1-ranked player. The Tim and Tom Gullikson Foundation was formed according to Tim's wish to help victims of brain tumors and their families by providing, care, psychological support, and research

Chapter 9

Medical Practitioners in the Animal Kingdom

Scarlet macaws of Central and South America often eat unripe fruits and nuts, which other animals avoid. Many may contain bitter-tasting alkaloids. Alkaloids are basic nitrogen compounds which are usually very bitter tasting and often poisonous. These birds also ingest clay from riverbanks. It has been suggested that the clay absorbs some of the compounds upsetting to the bird's digestive system.

Many animals seek salt (sodium chloride), essential to their well-being, but lacking in their diet. Moose often lick the salt from roads in wintertime. Animals are said to lick the salt off rocks left from evaporation of water from streams. In Kentucky, small streams are called licks and there is a town called Salt Lick. The sheep and deer on the Shetland Islands and the Isle of Rum in the Hebrides get minerals they need by biting off the legs of nesting chicks of seabirds.

Elephants have for two million years mined clay from an extinct volcano in Kenya called Mount Elgon. The elephants must negotiate difficult terrain to reach the cave as attested by skeletons along the route of elephants that didn't make it[20]. The elephants appear to be after salt, which is lacking in their herbivorous diet. In chapter 3, it was pointed out that when an elephant is on his last set of molars, he puts himself on a "soft-food diet." Is he acting as a dietician, or perhaps a dentist?

---

[20] One is reminded of Hannibal who brought elephants over the Alps to attack ancient Rome. His attack may have been called off because the elephants suffered illness from climbing through cold mountain passes.

Both mountain gorillas and chimpanzees eat clay as a remedy for upset stomachs. The chimps often get their clay by breaking off pieces from termite mounds. Similar clay called kaolin is sold in pharmacies for human stomach problems. Chimpanzees in Tanzania are known to eat Aspilia leaves which are rough and have sharp barbs and a disagreeable taste. It was later learned that local herbalists used these leaves to treat stomach upsets and to rid the gut of parasites. Examination of the chimp excreta revealed tiny worms on the barbs of the Aspilia leaves.

At the Awash Falls in Ethiopia, baboons living at the bottom of the falls are afflicted with parasites which come from the snails. The baboons treat themselves by eating the fruit of the *Balanites aegyptiaca* tree. The baboons at the top of the falls have access to the same trees but do not eat the fruit. There are no snails at the top of the falls.

Wild animals often seem to be healthier than similar domestic animals. Domestic sheep need worming seven times a year. Wild animals have fewer parasites and often suffer less from diseases that afflict them, though the same maladies are devastating to domestic animals. Wild deer with tuberculosis and wild boar with swine fever may show no signs of distress.

A male dolphin living at Marineland, Florida, shared a tank with his pregnant mate. The male dolphin was observed swimming up and down the body of the female, seemingly examining her closely. He became very agitated and, swimming rapidly, banged his head against the wall repeatedly. A short time later, the female gave birth to a stillborn. Dolphins are well known to be sensitive to sonic signals. Was he giving his mate a sonogram? It is tempting to believe that he was.

Chapter 10

Hierarchy in Various Animals

Hierarchy is established among young elephants during harmless sparring without the threat of injury. This protects the elephants while providing the advantages of hierarchy. Battles end with the weaker contender giving up because he knows his rank in the hierarchy so that no one gets hurt. Fights between elephants, while the potential for injury is present, rarely end up with serious wounds. Elephants could use their tusks to inflict mortal injuries, but rarely resort to such dangerous tactics. Usually the contest is settled based on strength and endurance and ends with one elephant giving up and walking away.

Fights between males over territory or in competition for mates occur in a wide variety of species. The mockingbird is quite territorial. While teaching chemistry at Morehead State University in Kentucky, I had an opportunity to observe the behavior of a mockingbird whose territory was the grounds of the house next door. The chemistry department was on the fourth floor overlooking the house where the bird spent time perched on the roof. My friend and colleague Tony Phillips, who used to teach the bird-watching course, pointed out this mockingbird and told me that it would fly down to the street level and fight with its own image in the side mirrors of the cars parked in front of the house. There was a large holly tree in the front yard, and the mockingbird liked to eat the red berries from this tree. During his vigorous efforts in the contest with the bird in the mirror, the mockingbird would defecate on the car door. On my frequent walks past the house, I often noticed the berry stains. Here in Jacksonville Beach where I now live, I have seen similar stains on cars. There are more mockingbirds in Florida than there are in Kentucky.

Male giraffes have spectacular fights over mating competitions. Two males will "square off" by standing side by side, facing in opposite directions.

Then one will swing his head in as large an arc as possible, colliding with the other's belly. The long neck makes the collision so violent that the giraffe receiving the blow may be knocked off his feet. The object seems to be to cause internal injury, which is possible, but rarely happens. Some say that these fights result in a lot of lumps and bumps on the heads of male giraffes.

Two hippopotamuses fighting

Hippopotamuses are territorial in water, but not on land. The hippo spends its days lolling restfully in the river (the fragment hippo—refers to "horse" and—potamus to "river") and nights foraging for grass on land. Males have fights over mates and territory and frequently inflict nasty wounds on each other. Hippos exude a liquid which hardens or dries on the skin and turns a reddish color. This substance is said to protect the hippo from sunburn and is also said to have antibiotic properties. The wounds endured by male hippos heal quickly and cleanly. It seems very probable that these wounds are exposed to plenty of pathogens as the hippo's way of marking his territory is to swirl his tail while he defecates, thereby spreading his feces as widely as possible in the shallow water.

Chapter 11

Amphibians

Amphibians include frogs, newts, and salamanders. The following discussion will be limited to frogs. Frogs are tail-less amphibians of the order Anura (from the Greek an-, meaning "not," and -oura, meaning "tail"). Frogs are ectothermic (cold-blooded), passively adjusting their temperature to that of their surroundings. Frogs are semiaquatic and have short bodies and long muscular hind legs, allowing them to move on land in spectacular leaps of twenty times their body length. The frog's neck is short and thick with little capability of turning. Nevertheless, the frog has a 180-degree field of vision due to its bulging eyes. Toads[21], though not popularly recognized as frogs, are indeed frogs. The popular notion of a toad is a terrestrial animal similar to a frog with a drier, warty skin. Toads also have shorter hind legs than do other frogs. The order Anura has numerous species, about five thousand in number. Frogs have had a great deal of time to diversify. They have been around at least as long ago as dinosaurs. Frogs range in size from 0.39 in. (10 mm) for the Cuban tree frog to 12 in. (300 mm) for the goliath frog of Cameroon, Africa.

Frogs have only one body cavity, which contains all of its internal organs. In contrast, the human has the chest, abdominal, and pelvic cavities.

---

[21] You may find it interesting to read of a toad mystery which occurred in Hamburg, Germany, in 2005. Over a thousand toads swelled up to three times the normal size. When they crawled out of the water, making screeching sounds, they suddenly exploded, scattering their innards about. It was established that crows were responsible for these toad explosions. The crows discovered that with a well-placed thrust of their bill to the toad's body cavity, they could extract the toad's liver. The toad's defense mechanism was to puff itself up to intimidate their predator, and the pressure that resulted forced their organs out of the wound.

In humans, the chest cavity is separated from the abdominal cavity by a muscular diaphragm, important in breathing. Frogs have neither a diaphragm nor ribs to aid in breathing. The frog breathes by lowering the floor of its mouth, causing the throat to puff out. Air enters through the nostrils. The nostrils close, and the floor of the mouth contracts, forcing the air into a pair of lungs. Much of a frog's respiration occurs through its skin. The skin is thin and permeable to oxygen, carbon dioxide, and water. Oxygen diffuses through the skin directly into capillaries close to the skin's surface. Carbon dioxide leaves the bloodstream by diffusing out through the skin. Respiration by use of the skin allows the frog to breathe underwater. One disadvantage of breathing through the skin is that the skin is permeable to environmental contaminants as well as oxygen and carbon dioxide. The world in general is becoming more polluted, and this may explain an alarming decline in frog populations.

Oxygen and nutrition are supplied to the frog's tissues, and carbon dioxide waste is cleared from the body by the circulatory system. Frogs have a three-chambered heart. Blood enters the heart by two upper chambers called auricles and leaves the heart from a single ventricle. Oxygen-poor blood from the frog's body enters the right auricle and passes down a vein to the bottom of the ventricle. Discharging oxygen-depleted blood at the bottom of the ventricle is crucial to the quality of the frog's circulation. Next, oxygen-rich blood from the lungs enters the left auricle and passes into the top of the ventricle. The oxygen-rich blood tends to stay at the top of the ventricle because the bottom is already filled with oxygen-poor blood. The ventricle contracts and the oxygen-poor blood is directed to the lungs where it will pick up oxygen and discharge carbon dioxide. The oxygen-rich blood will then be directed throughout the body. This arrangement minimizes but doesn't completely prevent the mixing of oxygen-rich and oxygen-poor blood even though they are both present in the ventricle at the same time. This gives the frog a richer supply of oxygen and allows a higher metabolic rate. As a result of this higher metabolic rate frogs have a higher body temperature. Frogs are also more active than their cousins, the newts and salamanders, in which oxygen-rich and oxygen-poor blood freely mix.

Frogs have large mouths and long sticky tongues which fold back toward the throat. The frog is very sensitive to motion, and when an insect passes by, the frog can suddenly flick its sticky tongue to capture the morsel. Frogs have teeth only in their upper jaws. Frogs also have teeth in the roof of their mouth called vomerine teeth. Frogs do not use their teeth for chewing, but only to hold prey until they get a better grip. Frogs pull their eyeballs downward, aiding in swallowing their prey whole.

Frogs are often nocturnal, and like most nocturnal animals, they possess a mirror behind the retina which reflects the light back through the retina for a second pass. This reflective layer is called the tapetum lucidum and occurs in dogs, cats, deer, and crocodilians and is especially prominent in nocturnal species. This produces an eye glow as we might see in a deer caught in the headlights. This second pass through the retina greatly (about 40%) increases light sensitivity at night. Humans do not have a tapetum, and the red-eye often seen in flash photography is due to blood.

The eye's ability to focus in air is mainly due to the refraction of light as it passes from air into the aqueous medium of the cornea. This is due to the large difference in density between air and water, which results in a big difference in the refractive index. The lens of the eye is located behind the cornea. Light entering the lens passes from one watery medium (cornea) to another (lens). There is only a small difference in refractive index and so leads to only minor focusing. Humans and many other animals can fine-tune the focusing of light by muscular action to change the curvature of the lens. Frogs and fish focus by a different strategy, that is, by moving the lens toward or away from the retina like the lens of a camera. When underwater, a frog is seeing light as it passes from water into the watery cornea of the eye, and so the frog loses the strong focusing action provided by vision in air. Frogs see better in air than underwater.

Another impediment to the frog's underwater vision is due to a protective membrane which closes over the eyes. This is the nictitating membrane[22] or "third eyelid," which occurs in many animals such as dogs, cats, birds, crocodilians, hippopotamuses, elephants, camels, and seals, but not humans.

Frogs do not have external ears. Their tympanic membrane, corresponding to our eardrum, is on the surface of their head. They use hearing to receive mating calls. They are able to perceive lower-frequency vibrations through their skin. The mating calls of male frogs are very loud and can be heard from a great distance. The calls of frogs are uttered with the mouth closed.

During mating, the male frog mounts the female and grips her body tightly in an embrace known as amplexus, which can last for hours or days. The female releases her eggs which are then fertilized by the male outside the female's body. The eggs swell and secrete a gelatinous coating. They are normally laid in water and have the appearance of a complex string of pearls. Frogs produce numerous eggs to compensate for many eaten by predators. A further hedge against predation is synchronization of breeding. Many frogs breeding at the same time leads to eggs and offspring so numerous that individuals are less likely to be devoured. Predators are overwhelmed by their numbers. Frog eggs hatch to produce tadpoles with eyes on the sides of

---

[22]   The nictitating membrane serves a unique function in woodpeckers. Besides protecting the eyes of the woodpecker from flying wood chips, the nictitating membrane holds the bird's eyes in its head as the bill slams into the trunk of a tree.

their heads. As metamorphosis occurs, the tadpole gradually changes into a frog with legs and without gills. The tail is lost by apoptosis (programmed cell death) and reabsorption.

Tree frogs are able to grip vertical surfaces by use of "toe pads." These toe pads enable the tree frog to cling to glass surfaces. When the frog presses down on the glass, water is driven out between groups of cells. The frog's foot is held to the glass by surface tension of the water. This small force is sufficient to support the weight of the frog. Surface tension of wet sand, not immersed in water helps to support the weight of automobiles. The Costa Rican flying tree frog extends the webbing between its toes to allow it to glide from branch to branch like a flying squirrel.

The wood frog lives north of the Arctic Circle. This species can survive with 65% of its body frozen. When solutions freeze, it is predominantly water that solidifies. This means that other dissolved components are concentrated in the solution that remains. The more concentrated solution has avoided freezing because the more concentrated the solution, the lower its freezing point. In the wood frog, glucose is the substance which concentrates as water freezes in the wood frog's body. The glucose acts as antifreeze.

The marsupial frog carries her eggs in a pouch the way marsupials carry their babies. When the eggs hatch, she pours the tadpoles into environmental water. In the Suriname toad of South America, the eggs gradually sink into the skin of the female's back. After a few days, the young frogs struggle free of these pockets. An Australian species called the gastric brooding frog swallows her fertilized eggs. The tadpoles take eight weeks to develop in her stomach and then hop out of her mouth. During tadpole development, stomach acid and digestive enzyme secretion is put on hold. In another species, known as Darwin's frog, the male swallows the eggs and the tadpoles develop in his vocal sac. During the first half of the twentieth century, the African clawed frog was used in a pregnancy test. The hormone, chorionic gonadotropin, occurs in a pregnant woman's urine. When the urine is injected into the female frog, the frog is induced to lay eggs, a positive result indicating pregnancy.

Chapter 12

Reptiles

Reptiles are exothermic and cannot produce bodily warmth by metabolism of their food. Reptiles, unlike amphibians, have a tough scaly skin. They lay eggs covered with a waterproof shell. Reptiles were the first animals to possess backbones and to be completely adapted to live on dry land. Reptiles have a cloaca through which solid and liquid wastes as well as sperm and eggs are discharged. Dinosaurs, which were reptiles, were the dominant creatures during the Mesozoic Era (248-265 million years ago.) Reptiles live everywhere except in very cold regions. There are nearly six thousand species of reptiles. Some reptiles discussed below are turtles, snakes, crocodilians, and lizards. Birds are discussed separately, though many scientists consider birds to be descended from dinosaurs and should therefore be considered to be reptiles.

Turtles[23] are exothermic (cold-blooded) reptiles, most having a cartilaginous shell protecting their body. Turtles are the most ancient of all vertebrates, having existed on earth for two hundred million years. The shell is developed on the turtle's ribs. Unlike other vertebrates, the shoulder and hip girdles of turtles are located inside the rib cage. Turtles are divided into two groups: Cryptodira and the Pleurodira. The Cryptodira draw their head into their shell by bending their flexible neck into an S shape in the vertical plane. The Pleurodira partially retract their head by bending it in the horizontal plane under the shell. The Pleurodira are restricted in their range to South America, Africa, and Australia.

---

[23]   An interesting observation concerning turtles is the absence of degradation of tissues with increasing age. A one-hundred-year-old turtle has muscle, heart, and liver tissue in as good a condition as that of a young turtle. This is certainly not true of humans.

Turtles have lungs located toward the top of the shell, and they must have access to air to breathe. A turtle's ribs are attached to the upper shell or carapace and so are not involved in breathing. The bottom shell is called the plastron. Turtles have two sets of muscles used in breathing. One set stretches the body outward from the shell, thereby expanding the body cavity and allowing the turtle to inhale. The other set draws the body inward in order to exhale. Turtles in water are able to extract some oxygen directly from the water by tissues in the back of their mouth, so-called pharyngeal respiration. This extends the time that the turtle can remain submerged. Some turtles have tissues in their cloaca which are also permeable to oxygen. The cloaca in these turtles has a dense bed of capillaries to aid in transport of oxygen and carbon dioxide.

The Fitzroy River turtle discovered in 1973 lives only in the Fitzroy River in Queensland, Australia. This turtle gets 68% of its oxygen through its cloaca and rarely needs to come up for air. The Fitzroy River is rapid flowing and well oxygenated. Turtles have extraordinary capacity to survive without oxygen. They have lived up to 33 hours in a 100% nitrogen atmosphere. Other reptiles have greater capacity to live without oxygen as compared to mammals, but they can only survive without oxygen for about 30 minutes.

Soft-shelled turtles and the leatherback sea turtle are able to absorb oxygen from water through their shells. This has the disadvantage that chemical contaminants in the water are also absorbed into the turtle's bloodstream. Sea turtles are able to shed salty tears to rid the system of salt from drinking ocean water.

Some turtles, such as the Galápagos and box turtle are predominantly land animals. They have eyes looking downward at the ground. Turtles that spend most of their time in water, such as the snapping turtle and the soft-shelled turtle, have eyes at the top of their head. They are able to submerge their bodies with only their nostrils and eyes projecting above the surface.

Turtles have good night vision, having eyes with many rod cells in the retina. The turtle eye is also supplied with many cone cells used for discerning color. Turtles can see the light spectrum from the near ultraviolet (invisible to humans) to the red. Turtles are famous for their slow locomotion. However, carnivorous turtles like the snapping turtle are able to move their heads quickly to seize their prey. Turtles cannot stick their tongues out to seize their food the way many other reptiles can.

The leatherback turtle is the largest living turtle in the world. Leatherbacks weigh up to 2,000 lb. (900 kg). The large size of this turtle results in a low surface-to-volume ratio which aids in the retention of bodily warmth. When first leatherbacks were found in the frigid waters off Nova Scotia, it was thought that they were strays and that this must be a rare freak occurrence.

Further observation confirmed that this was, in fact, a standard environment for these turtles. Leatherbacks migrate to the Caribbean in order to lay eggs on warm, sandy beaches. Leatherbacks tend to lay their eggs on the same beach where they were born. Some researchers have suggested that the leatherback's ability to withstand cold temperatures is due simply to its huge size and have proposed to classify it as a "gigantotherm." Others have noted that the

Leatherback turtle

leatherback has large deposits of brown adipose tissue (BAT) which permits the generation of heat and gives the leatherback turtle a degree of endothermic character.

The green sea turtle is able to maintain a constant heart rate over a temperature range from 54.5 to 77.9°F (12.5-25.5°C.) This is an indication that green sea turtles are not purely exothermic. Sea turtles in general are more active than other reptiles. Recall that a by-product of activity is production of heat. Sea turtles have enough myoglobin, an oxygen-carrying pigment, in their blood to support aerobic metabolism. Aerobic metabolism (requiring the presence of oxygen) is characteristic of endothermy. Sea turtle muscles are reddish in color similar to birds, which are clearly endothermic. BAT is helpful in maintaining nearly constant body temperature in young altricial mammals, but animals such as birds manage well without it.

Snakes are carnivorous reptiles without legs. They have teeth but cannot chew their food and swallow their prey whole. Their jaws are flexible, and the upper and lower jaws can separate, permitting them to swallow prey larger than the width of the snake. Snakes vary in length from the 4 in. (10 cm) thread snake to the 25 ft. (7.6 m) anaconda and python.

Snakes have no visible ear canal and have a single bone in the middle ear, in contrast to mammals that have three auditory ossicles. Hearing in snakes is poor, but they can sense vibrations through the skin of their belly. Snakes have no eyelids and cannot move their eyes. Acuity of vision varies widely with the species. Some tree snakes see quite well, while some burrowing snakes can only tell dark from light. Pit vipers (rattlesnakes and pythons) are able to detect infrared radiation, permitting them to see warm or cold objects in the dark. Pit vipers have special organs located in pits in front of their eyes that are sensitive to infrared radiation. Infrared radiation is constantly emitted by any warm object. Snakes focus their eyes by moving the lens nearer or farther from the retina as with the lens of a camera. This mode of focusing is shared with fish and turtles. The pupils of rattlesnake eyes are oval, whereas the pupils of nonvenomous snakes are round.

Snakes shed their skin (molt) as they grow, but continue to shed their skin, though less frequently, even after fully grown. Whether this is due to slower growth or whether shed for another reason is debated. The skin is usually peeled away in one piece, being turned inside out from the head to tail. This renewal leaves the snake with a brightly colored new skin. The molting of skin has led the snake's association with healing and is used in a symbol of the medical profession, the "rod of Asclepius," which depicts a snake wound around a rod. The winged rod with two snakes is the double caduceus often erroneously used as a symbol of medicine.

Most snakes, said to be oviparous, lay eggs covered with a leathery waterproof shell inside which embryos develop outside the mother's body. Snake eggs are not incubated by the adult snakes, as are bird eggs. Snakes, being exothermic, cannot produce bodily warmth to incubate eggs. An exception is the python that is able to produce warmth by rhythmic body contractions. This is very unusual in exothermic animals and is analogous to heat production by shivering in mammals. Examples of oviparous snakes are bull snakes, milk snakes, and green snakes. When the young are fully developed, they break out of their shells using an egg tooth like that of birds. Other snakes, described as viviparous, undergo pregnancy with young attached to a placenta inside the mother. Ovoviviparous snakes retain eggs inside the mother's body until the young snakes are fully developed. The young emerge either within the mother's body or emerge from the egg immediately after the latter is laid. It is often difficult to tell whether a snake is viviparous or ovoviviparous. Some authorities say the common garter snake (*Thamnophis sirtalis*) is ovoviviparous; others claim this snake is viviparous.

Snakes have only one lung, presumably to save space in their narrow body. Their kidneys are located one ahead of the other. The snake's heart can move over to make room for large meals which the snake must swallow whole. The garter snake can live as far north as within a few hundred miles of the Arctic Circle. The common adder, also called the European viper, ranges north of Arctic Circle in Norway. This is farther north than the range of any other snake. Snakes generally prefer warm climates. Garter snakes prey mostly on earthworms and amphibians. The latter seem to be sensitive to the slightly toxic bite of the garter snake.

Snakes have a three-chambered heart so that the major source of pumping action, a single muscular ventricle, is used to send blood to the lung to pick up oxygen and discharge carbon dioxide. The same ventricle must then send the purified blood to the rest of the snake's body. The oxygenated blood and the spent blood must mix in the same ventricle. The snake is not getting good, clean oxygenated blood to his body. Snakes are fast when they strike, but they can't keep up a sustained effort for very long.

Rattlesnakes are venomous and have hollow fangs at the front of their mouths for injecting venom into their prey. Rattlesnake venom breaks down blood cells and dissolves tissues. Rattlesnakes are able to adjust the amount of venom according to the amount needed for prey of various sizes. Often the rattlesnake after striking its prey waits for the animal to weaken or die before consuming it. An advantage to this strategy, especially for a relatively large animal, is that the venom will be distributed throughout the prey's body, partially digesting its flesh.

Rattlesnake ready to strike

Another advantage to waiting is to protect the fragile body of the snake from injuries inflicted by a struggling victim. Coyotes are known to violently shake rattlesnakes, breaking the snake's spine. Poisonous snakes are immune to their own venom. They also have reduced sensitivity to the venoms of other poisonous snakes.

Antivenins are produced by injecting horses with venoms from various venomous snakes in increasing doses. A mixed antivenin protects victims from the bites of all snakes in the United States including the coral snake. Coral snake venom is a neurotoxin. Persons allergic to horses cannot be treated with these antivenins. The more poisonous snakes of Australia, India, and South Africa can be countered by antivenins in the same way, except that these antivenins are effective only for one species at a time.

The Indian cobra has been a favorite snake for snake charmers for centuries. When disturbed, the snake rises and spreads its ribs to form a hood. The snake sways as the snake charmer plays his flute and moves the instrument from side to side. The snake also sways in rhythm with the motion of the flute. The snake cannot hear and is responding to the flute's motion. Sometimes the snake charmer protects himself by removing the cobra's fangs. Since 1972, it has been illegal to trap snakes in India. Recently this law has been enforced, and snake charming is becoming a lost profession. Cobra venom is a neurotoxin.

The flowerpot snake, usually called the Brahminy blind snake, native to Southeast Asia, was introduced into Florida in the soil of imported plants. It has similarly been introduced into several other tropical places. This snake inhabits soil in gardens. The average size of adults is 2.5-6.5 in. (6.4-16 cm). The blind eyes are tiny dots under scales. This snake is dark gray and resembles an earthworm, but is not segmented. The flowerpot snake feeds on ants and termites in all of their life stages. It may lay eggs or may bear live young. All individuals are female. Eggs divide to produce new individuals without involvement of sperm. The offspring are all genetically identical and may be up to eight in number.

Crocodilians include crocodiles and alligators. Alligators include the American alligator (*Alligator mississippiensis*); the -ensis ending means belonging to. There is a Chinese alligator (*A. sinensis*); the Latin *sinae* means China. Crocodiles include the saltwater crocodile, Nile crocodile, caiman, and the gavial. There are approximately twenty-three species of crocodilians. These carnivores are equipped with powerful jaws and sharp conical teeth. The long jaws can close on prey animals with great force, but the closed mouth can be held closed with relatively little force. Crocodilians are "sit and wait" hunters. They lie submerged at water's edge until prey come to take a drink or cross a river as wildebeests in migration. Then they lunge with great speed, grasp the hapless prey, and drag it with irresistible force underwater where it drowns. The crocodile or alligator then grasps a chunk of flesh or a limb and, with a violently rolling motion, tears the flesh or limb asunder and swallows it whole. Crocodilians have numerous pointed, cone-shaped teeth for tightly grasping their food, but they cannot chew.

Crocodilians have a four-chambered heart, nearly like mammals. They didn't quite make it to four separate chambers since there is a hole in the wall between the left and right ventricles, allowing freshly oxygenated blood to partially mix with deoxygenated blood.

The American alligator is easily distinguished from the crocodile by its broad snout and by the fact that its lower front teeth fit into sockets in the upper jaw when the mouth is closed. The lower teeth of a crocodile protrude from the mouth when closed. An alligator, if forced onto its back, seems to go into a sort of trance, becoming immobile for a time[24]. It may be that the alligator is merely disoriented.

None of the crocodilians have sex chromosomes. The sex of an alligator is determined by the prevailing incubation temperature before the gator hatched. More specifically, it is during the first half of the incubation period that temperature exerts its influence on sex determination. In Louisiana, it was found that 60-80% of wild gators sampled were male, but of gator eggs incubated artificially, only 10-25% produced males. The eggs were cooler in the artificial environment. Low temperatures in the 82°-86°F (28°-30°C) resulted in females. Temperatures in the higher 90°-93°F (32°-34°C) produced

---

[24]   A similar phenomenon occurs with chickens grasped by their legs and held upside down. No chickens were harmed by this procedure. I'm not sure that the same can be said about the men turning gators on their backs.

males. In intermediate temperatures, both males and females in various proportions were the result.

It is time to take the unusual opportunity to confirm a centuries-old myth. Crocodiles were said to shed tears while consuming their prey. The phrase "crying crocodile tears" means shedding tears of regret with no true contrition whatsoever. Crocodiles are merciless when tearing their prey limb from limb in preparation to swallowing the pieces whole. Crocodiles do not truly regret having violently killed their prey. Even Shakespeare refers obliquely to crocodile tears in his play *Othello*, act 4, scene 1:

> O devil, devil!
> If that the earth could teem with woman's tears,
> Each drop she falls would prove a crocodile.
> Out of my sight!

Dr. Kent Vliet, a zoologist at the University of Florida, Gainsville, explains and illustrates in a video that alligators and crocodiles do shed tears while they eat. You can view this video at http://videos.howstuffworks.com/university-of-florida/3800-crocodile-tears-are-real-video.htm. According to Dr. Vliet, the musculature involved in working the jaws of a crocodilian are also associated with its eyes.

Chapter 13

Birds (Aves)

Birds are descendants of reptiles. In fact in the modern cladistic approach to classification developed since 1960, birds are considered to be avian reptiles. What are considered classically to be reptiles are called nonavian reptiles. The cladistic method relies on evolutionary relationships of animals and is supported mainly by molecular evidence. Both protein and DNA evidence is used these efforts. In a 2008 report in the journal Science, protein residues from a sixty-eight-million-year-old *Tyrannosaurus rex* femur bone was compared with proteins from chickens, ostriches, and alligators. Only six peptides (similar to proteins, but smaller molecules), made up of eighty-nine amino acids, were available from the *T. rex* sample. This was enough to support the conclusion that the dinosaur was more closely related to birds than to alligators.

Birds have scales (on their feet) and lay eggs like other reptiles. Unlike other reptiles, birds are endothermic (warm-blooded). Whether or not some dinosaurs were endothermic is debated. Birds have feathers; cladists say that feathers are specialized scales. By chemical manipulation to activate genes, it has been possible to get certain chicken scales on legs to produce progeny with feathers in place of scales. This suggests that feathers and scales are closely related.

Most birds fly. Those that fly have bones filled with air, which lightens their bodies for flight. Flightless birds such as ostriches, emus, and penguins have marrow-filled bones. Penguins need a higher body density to help them dive deep underwater. The solid bones of the penguin aid the bird in diving. Though flying birds have bones filled with air, there is still enough bone marrow in adult birds to generate sufficient red and white blood cells. The thighbones of flying birds are filled with marrow.

Mammals have lungs and use a diaphragm to actively inhale and passively exhale. In birds, exhalation is active and inhalation is passive. In mammals, air fills the lungs and mixes with stale air left over from the previous exhalation, which is never 100% complete. Birds do not have saclike lungs as mammals do and have no diaphragm to help them breathe. Birds use muscles to lift the breastbone outward to inhale. During inhalation, air is directed to four air sacs toward the rear of the body. Next, the bird exhales, with more force than inhalation, driving air into a multibranching system ending in a bed of very fine tubules (air capillaries). This bed of tubules corresponds to the mammalian lung. It is in these tubules that oxygen and carbon dioxide are exchanged with the bloodstream. When the bird next inhales, air is driven from the air capillaries into three symmetrically placed forward air sacs. Finally, during a second exhalation, air leaves the bird's body through the mouth. It requires two inhalations and two exhalations to move one breath of fresh air in and out of the bird's respiratory system.

The result of this arrangement is that the avian "lung" is continually filled with fresh air, not air mixed with stale air left from the previous breath. This air contains a higher percentage of oxygen needed to supply the higher metabolic rate of a bird as compared to that of a mammal. Birds have a higher body temperature, 104°F (40°C), compared to 98°F (37°C) for mammals. Bird flight requires a higher rate of energy expenditure than that which mammals use in moving. However, birds more than compensate by arriving at their destination in much less time. Still, during their briefer period of flight, they need more oxygen.

The air sacs used by birds in respiration extend into their bones and are in contact with their viscera (stomach, liver, intestines). I can't say "in contact with their abdominal cavity" because there is no such thing. In birds, there is no diaphragm to create separate chest and abdominal cavities. As a result, respiratory infections in birds can spread into the viscera and into bones. Birds have a similar vulnerability to pollutants. It was no accident that coal miners took canaries[25] with them into the mines.

The circulatory system in birds is superior to that found in nonavian reptiles. Unlike reptiles, birds have a four-chambered heart, just as humans do. Oxygen-depleted blood from the bird's body is sent to the "lungs" where carbon dioxide is released from hemoglobin and oxygen is bound to hemoglobin. The oxygenated blood is then sent to the left side of the heart and pumped

---

[25] Canaries were named for the Canary Islands where they originated. In AD 60, King Juba of Mauritania sent an expedition to the "Fortunate Islands" where they encountered large ferocious dogs on one of these islands. They named the island Canaria from the Latin canis for "dog." Indirectly, the canary birds owe their name to the dogs. There is a modern breed of dog called the presa canario, bred by the Spanish in the Canary Islands.

to the body again. The oxygenated blood does not mix with the deoxygenated blood as happens with the three-chambered reptilian heart. One advantage that the avian circulation has compared to the mammalian system is that blood returns from the "lungs" by four rather than two pulmonary veins. This provides the birds with more output of oxygen-laden blood. Birds need better circulation because flapping wings consumes a great deal of energy. The red blood cells of birds are nucleated. Mammalian red cells have no nuclei.

Birds possess keen visual and auditory senses. The mating behavior of birds involves songs and visual plumage displays. The sense of smell is of little importance to birds. In fact, some authorities used to argue that this sense was absent in birds. Now we know that turkey vultures find rotting carrion mainly by smell. It is significant that the great horned owl is the only predator of the common skunk. Mammals are more strongly attuned to smell than to vision. Dogs recognize humans and each other principally by odor. Mammals depend on pheromones detected by the Jacobson's organ to aid them in mating behavior.

Vocalization in birds is an important part of their life. We are all familiar with the use of song in mating rituals of birds. The complex singing may be aimed at staking out a claim for nesting territory. The singing may be telling competitors, "Don't build a nest here, this area is taken." The fewer birds in the area, the easier it will be to feed the young. Birds have a larynx, but it is vestigial and not used for vocalization. At the bottom of a bird's trachea, just ahead of the two bronchial tubes, is an organ called the syrinx[26], which acts as the voice box. Birds also call to warn others of the approach of predators and use voice for social communication. Turkey vultures can only make a grunting sound. They have no syrinx.

If birds sing, they must be able to hear. Hearing in birds is acute. The bird's range of perceptible frequencies is somewhat broader than that of a human (2,000 to 4,000 cycles/sec). There are differences in hearing between humans and birds. Humans process sounds in bytes of 1/20 second duration. Birds can discriminate sounds of 1/200 second duration. As a consequence, while we hear just one note, a bird might hear up to ten notes from the same source. Birds have perfect pitch, something shared by very few people. Most humans have relative pitch. We can recognize a song regardless of the key in which it is sung. This is not the case with birds; change the key, and the bird is lost. Some birds are exquisitely sensitive to soft sounds. Owls not only have very keen hearing, they also have the ears at slightly different levels on

---

[26]   In Greek mythology, Syrinx was a nymph pursued by the amorous Pan. The river nymphs tried to help her avoid Pan by changing her into hollow reeds at the river's edge. Pan fashioned the reeds into the pan flute. The French composer Claude Debussy composed "Syrinx (La Flute de Pan)".

their heads. This gives them the ability to accurately determine the location of the sound source. A barn owl can locate a squeaking mouse and capture it in total darkness. A snowy owl can locate and pounce on a mouse tunneling under the snow.

Birds begin to chirp while still inside the egg. They also hear and learn to recognize their mother's voice while still in the egg. When the chick is ready to leave the egg, it breaks the shell using a sharp egg tooth on the tip of its bill. There are special muscles in their neck to give them the strength to break the eggshell. Later, the egg tooth is discarded, and the special neck muscles become smaller. Predators also listen to the chick's chirping inside the egg and may attack and devour the chick. In the banded gull, so called for the black band near the tip of its bill, sound perceived by a chick inside the egg is very important. Chicks incubated artificially are deprived of auditory input from their mothers. Normal gull chicks peck at their mother's bill, which stimulates her to regurgitate food for the infant. Chicks incubated in silence fail to peck at the mother's bill and may starve.

Birds have excellent vision. Generally, birds have little or no ability to move their eyes within their heads. Their eyes are often toward the sides of their heads with limited binocular vision which depends on overlap of visual fields of right and left eyes. In humans, binocular vision covers 140 degrees of a total visual field of 180 degrees. A pigeon has an overlap 20 or 30 degrees of a total field of 300 to 340 degrees. Birds of prey need good binocular vision in order to catch their prey with an accurate strike. Birds of prey have their eyes placed looking forward and have good binocular vision. Owls cannot move their eyes at all and have a total visual field of 110 degrees with a binocular field of 70 degrees. They must turn their heads to see objects toward the side and are able to swivel their heads through 200 degrees.

Herbst corpuscles in joints or beaks of birds are exquisitely sensitive to vibrations. This sensitivity enables parrots to sense approaching earthquakes before humans have a clue of danger. Numerous parrot owners report that their birds all became agitated fifteen minutes before a particular quake became evident to humans. Herbst corpuscles are believed to help roosting birds sense a predator climbing the tree where they are at rest.

Birds have sex chromosomes like humans. In humans, the female is homogametic (XX), and the male is heterogametic (XY). During the production of sperm or egg cells (gametes), half of a chromosome pair is used to make the sperm or egg. Thus, a given sperm can be either X or Y, but an egg can only be X. If, during fertilization, the sperm is X, the offspring will be female; if the sperm is Y, the offspring will by male. In birds, the system is similar except that the female is heterogametic (ZW) while the male is homogametic (ZZ). W and Z are used to distinguish this system from the mammalian X and

Y system. Thus, in humans, the male determines the sex of the offspring, but in birds the female determines its sex.

Semen is transferred to the female during a brief (fraction of a second) "cloacal kiss" in which the male everts his cloaca and brings it in contact with the cloaca of the female. Semen is stored until the female ovulates. Male birds do not normally possess a penis, though ducks and geese have a penislike organ. Female birds at the time of hatching have left and right ovaries and oviducts, but after hatching, the right ovary gradually degenerates so

Turkey vulture in heraldic pose

that only the left ovary is functional. Some birds like the crow are determinate layers; that is, they lay a definite number of eggs. Others, chickens and ducks, are indeterminate layers; they lay an additional egg when one is removed from the nest. This tremendously increases egg production in chicken farms. This "double clutching" was very useful in the captive breeding program to restore the California condor population.

The turkey vulture is often called in error a turkey buzzard. The term buzzard properly refers to hawks whose soaring flight is somewhat similar to that of a vulture. English pioneers probably called all birds with similar flight patterns buzzards. In Hinckley, Ohio, a suburb of Cleveland, turkey vultures are migratory and are said to return to Hinckley on March 15. The first Sunday after March 15 is celebrated as "buzzard Sunday." Turkey vultures, like other vultures of the New World, have weak feet similar to those of chickens. Their feet are not useful for carrying prey but can be used to hold down carrion while they rend it with their beaks. Old World vultures, on the other hand, have strong feet with sharp talons. Vultures in the American continents have been shown by DNA evidence to have ancestry common to storks, flamingos, and ibises. The flesh of the turkey vulture's head is red like a turkey's and almost devoid of feathers. The nearly bald head is typical of vultures and prevents excessive contamination of their head with bacteria-laden flesh while dining. After poking around in a carcass, vultures bask in the sun with wings outspread to bake detritus and kill bacteria. This sunning posture is the so-called heraldic pose.

Turkey vultures, when disturbed on the ground, may spontaneously vomit their food. Authorities deny that they projectile vomit as is sometimes claimed. This may deter a perceived threat. It may be that after eating too much, the vulture needs to lighten his load so that he will be able to "take off" in order to avoid a perceived threat. Taking flight from ground level is difficult for a turkey vulture. He has to get a running start. The vomit has a foul odor and is strongly acidic. The gastric secretion of the turkey vulture is sufficiently acidic to kill noxious bacteria in the decaying flesh which vultures seek as food. Vultures eat only animals which are already dead and prefer to dine on carrion already somewhat decayed. The state of decay makes it easier for the bird to tear bits of flesh away with their beaks. The turkey vulture's digestive system is capable of killing bacteria causing anthrax, cholera, and botulism.

Vultures also defecate on their feet in order to cool themselves as they glide through the air. Some sources say vultures urinate on their feet. Not so, birds don't urinate. Whereas humans produce urea and excrete it in urine, birds save on water by ridding the body of nitrogenous waste as uric acid which is added to feces. The next time someone asks you, "What's that white stuff on top of chicken shit?" The answer is not, "More chicken shit." Tell them it's just uric acid.

It is now generally agreed that turkey vultures find carrion by both keen eyesight and by smell. That smell that was involved in the turkey vulture's quest for dinner was shown by the bird's attraction to underground gas pipelines after adding extra methyl mercaptan to the gas. Methyl mercaptan is a very foul-smelling, gaseous sulfur compound. This evidence is strengthened by the finding that methyl mercaptan is found in intestinal gas after a meal containing protein[27]. Engineers used this as a strategy for locating leaks in the pipeline. Where turkey vultures congregated, there would be the leak. If decaying carrion is hidden in brush, turkey vultures have no problem finding it. Not all vultures use odor to locate food. Vultures eat dead animals. They do not circle dying animals. Apparently, they have no premonition of impending death. Turkey vultures occasionally eat water plants.

It is often declared in literature that birds in general have little or no sense of smell, the turkey vulture being an exception. Some authorities dispute this, claiming that most birds can smell to some degree. The great horned owl is the skunk's only predator.

Vultures like to roost high in trees or on buildings. Sometimes, large groups of vultures, especially black vultures, roost for the night on tall buildings. They do not take flight until the sun has warmed the earth enough

---

[27] Sulfur itself does not have a foul odor as people often think. Well-water sometimes contains tiny amounts of dissolved hydrogen sulfide gas, H_2S. Hydrogen sulfide is responsible for the odor of rotten eggs. It is chemically related to H_2O, water. Sulfur is directly below oxygen in the periodic table of the elements. Hydrogen sulfide is extremely poisonous and often leads to fatalities associated with sewer gas. Frequently there is a double fatality when a rescuer is overcome trying to help a coworker.

to set up rising currents of air (thermals). The six-pound turkey vulture can ride on these air currents for hours without having to beat its wings.

The retina of the eye is equipped with rods, important for night vision and motion detection, and cones, important for color vision. Rods are prominent in nocturnal animals where colors are of little use. Cones are more useful to diurnal animals. Vultures have about one million rods and cones per square millimeter in the most sensitive area of the retina (fovea centralis). Humans have one-fifth as many in the corresponding region of the human retina. Vultures have distance vision about six to eight times better than ours.

Turkey vultures lay eggs (usually two) on the ground in caves, hollow logs, abandoned barns, or sheds. The turkey vulture doesn't build a nest. Both parents incubate the eggs. Like most eggs laid on the ground or in nests on the ground, turkey vulture eggs have a rusty, blotched appearance.

The hoatzin (*Opisthocomus hoatzin*) is a strange bird living in swamps and mangrove stands in the Amazon region. Its relationship to other birds is so controversial that it is at least temporarily assigned its own family and genus. Recent DNA evidence has shed little light on the proper classification of this unusual bird. The hoatzin (pronounced wat'son) has an unfeathered blue face with maroon eyes. It has a prominent crest and is 25 in. (63 cm) long. Its wings are rust colored, and its back is gray.

Two hoatzins on a tree bramch

The hoatzin's diet consists principally of leaves and a little fruit. As is clear from previous discussions above, this is a diet very low in nutrition and difficult to digest. The hoatzin relies on bacterial fermentation which takes place mainly in the bird's crop. The crop is an enlarged sac in the lower end of the esophagus. The crop is quite large in order to accommodate the large amount of leaves the bird needs to eat since there is so little food value to this diet. The crop is so large that the bird steadies itself by resting the crop's leathery patch on a tree branch. The large size of the hoatzin's crop does not leave enough room for normal musculature of the breast. For this reason, the hoatzin is a poor flier. The bacterial fermentation produces a fecal odor, and natives call the bird "stink bird."

Juvenile hoatzins have a claw on each wing. When they happen accidentally to fall from their nest into the water, they are able to claw their way up the tree and back to the nest. This is reminiscent of dinosaurs. Perhaps birds still retain a dinosaur gene which is activated in the hoatzin, but not in other birds.

The emperor penguin is the largest of all penguins, standing at about 3.75 ft. (115 cm) and weighing 48-81 lb. (22-37 kg). Their weight varies with the season; as when they are in the breeding season, they fast while staying at the rookery. The sexes are visually indistinguishable, though females are somewhat smaller than males. This penguin lives on the sea ice of the Antarctic and in the

Emperor penguin with chick

ocean surrounding the Antarctic continent. Adults are strikingly colored, black dorsally and white ventrally. They have ear patches of golden orange fading to light yellow toward and including the upper portion of the breast. The young have downy white feathers on the breast and black feathers on part of the head and the back. The ear patches of the young are white but become more yellowish with age. The young have a white chin, while adults have a black chin.

Males leave the sea in late March and begin marching toward a rookery, that is, a location where chicks will be raised. There is no food available on the ice, so once this march begins, the male penguins fast. The winter season in Antarctica begins in March and lasts until December. The breeding season is in the winter for emperor penguins. The temperatures may drop to -80°F (-62°C), and winds may blow at up to 100 mi./hr (160 km/hr). The rookery is located about 60 mi. (96 km) from the sea near an ice shelf, which offers some protection from severe winds. Males waddle 60 miles to the rookery. To rest from marching, the penguins lie on their belly and slide on the ice, pushing themselves along with their feet. Once at the rookery, they engage in a single pursuit, staying alive in the harshest winter on the face of the earth. To take maximum advantage of body heat, the entire mass of hundreds of male penguins stand shoulder to shoulder in a tight formation referred to by the French word for "turtle," tortue. Sometimes, the colloquial British term scrum[28] is used to describe this formation. The temperature is warmest in the center of the tortue and much colder at the perimeter. Penguins do not

---

[28]   Scrum or scrummage is a formation in the game rugby, in which opposing teams tightly interlock arms on shoulders. The two forwards try to gain control of the ball in the center to restart the game.

jostle to gain individual advantage. There is a very slow spiral motion of the entire tortue whereby each penguin will have his turn at all positions in the formation. Why do penguins choose the middle of winter to breed? And why do they choose an area for their rookery so far from the sea, which is their only food source? Part of the reason is that when the young penguins take to the water in the summer, there will be a plentiful supply of food for them. Another reason is that in the spring when the ice melts, they don't want their chicks falling through cracks in the ice. The ocean is much closer to the rookery in the springtime.

After some time, the female penguins arrive, and courtship and mating occurs. During a single breeding season, emperor penguins are monogamous. New mates are usually chosen for the succeeding breeding season, but sometimes last season's mating persists. The ratio of males to females is roughly two to three. The relative paucity of males is probably due to death of males trying to reach the sea following the hatching of chicks. (See below.) Courtships last for two or three weeks. In courtship, males drop their heads to their chest and utter calls interrupted by brief periods of silence. Females respond in a similar manner. If the female is satisfied with the pairing, they stay together, if not, she moves on to seek another male. The calls are highly individual and probably provide a mechanism for recognition.

Gestation of a single egg requires about sixty-three days, and a single egg is laid in May or early June, deep in the winter season. The egg is soon passed to the male that is solely responsible for incubation. The penguin pair practices the transfer of the egg from female to male. Sometimes, perhaps with an inexperienced pair, the egg rolls onto the ice and quickly freezes. Both parents retreat to the ocean, aborting their mission. The female's resources have been strained by producing the large 1-lb. egg (450 g), and she leaves to travel to the open ocean to forage for food.

Emperor penguins do not build a nest. There would not be any nesting materials in the area from which to build one even if they tried. The male places the egg on his feet and puts his weight slightly on his heels to keep the egg as far as possible off the ice. The male must balance this egg on his feet for the next sixty-three days in the darkness of the Antarctic winter, and sometimes in 100 mi./hr (160 km/hr) winds. He must also shuffle in and out of the tortue to stay warm and conform to penguin etiquette. A flap of skin hangs down and covers the precious egg.

The male has a breeding pouch in which the chick can stay warm. The pouch is a fold of skin bereft of feathers inside for better heat transfer to the chick. Females often return within a day or two of the egg's hatching. If the female does not return promptly, the male feeds the chick with an esophageal liquid, much in the manner of pigeons or flamingos, vide infra. He may need to feed the chick for up to ten days. When the female arrives, she feeds the chick and her mate by regurgitating food carried in a special pouch in her throat. This is the first time the male has eaten in about 115 days, and he has lost up to one-third of his body weight. The male now makes his 60 mile journey to the open ocean to forage. In his weakened condition, the male may just not make it all the way. Males perishing on this march may account for the disparity in male to female ratio.

The male and female take turns at foraging and caring for their chick. As summer approaches and open water comes closer with the melting of ice, foraging trips become more frequent. Soon the chicks can be left alone. The chicks form dense groups of their own called crèches. At an age of about five months, chicks leave the rookery and travel to the open ocean. By the time they reach open water in December or early January, they have molted their downy feathers and grown oily outer feathers. The oily waterproof feathers are needed to allow them to swim and gather food on their own. Food is abundant in this season and presumably helps to explain why emperor penguins choose to breed in midwinter.

The emperor penguin feeds on fish, squid, and crustaceans. Most penguins feed on krill in surface waters. The emperor penguin hunts by deep dives of up to 1,300-1,500 ft. (400-450 m). This is a deeper dive than that made by any other penguin. The smaller king penguin which resembles the emperor, except in size, also makes deep dives in search of food. It is not known how the emperor penguin finds its food at such depths where light does not penetrate. Northern fur seals can find food by using their very sensitive whiskers, but emperor penguins do not possess whiskers. Emperor penguins, like most other penguins, drink saltwater. Penguins have a supraorbital (above the eye) gland which extracts a highly concentrated salt solution from the blood. This strong saline solution is discharged through the bird's beak.

Emperor penguins are preyed upon by leopard seals and by orcas. The leopard seal has large sharp teeth. The seal waits at the edge of the ice for a penguin to dive in. The seal grasps the penguin by the feet. The seal then thrashes the penguin against the water's surface left and right repeatedly in order to skin the penguin. The seal then devours the carcass without having to eat the feathers. Orcas are known to treat seal pups in similar fashion in order to skin the animal before eating it. (Whether orcas treat penguins in the same way, the author has not been able to establish.)

Kingfishers do indeed feed mainly on fish, such as the belted kingfisher, the only kingfisher common in the United States. There are two other rare North American species, the ringed and the green kingfishers. Many kingfishers in Australia, Africa, and Asia do not live near water and feed on insects, snakes, and small animals. The male belted kingfisher is blue above with a white belly and a blue belt around the lower neck and is about the size of a pigeon. The female has an additional rust-colored belt at the top of the breast. Both have a jagged, bifurcated crest. The kingfisher is a strong flyer and may be seen hovering over clear water before diving for a fish. The ability to hover is not possible for most birds. The kingfisher may take the fish to a sturdy perch and bash the fish against the branch in order to subdue it and perhaps break its bones to aid in swallowing. The bird then tosses the fish in the air and catches it headfirst so that he can swallow it with the grain of the scales and fins.

The belted kingfisher (*Megaceryle alcyon*[29]) digs a burrow in a vertical sandy clay bank of a river or stream. The burrow points upward to prevent possible flooding and ends in an enlarged nesting chamber. Kingfishers have four toes, one pointing backward, and three forward. The two inner toes that point forward are fused (syndactile) for most of their length and act as a shovel during digging of the burrow. The burrow may be shared with swallows, which excavate a side pocket for their nest site. This is reminiscent of the sharing of an armadillo's burrow, vide infra.

The female lays six or seven eggs. Both parents participate in incubation, though the female incubates at night. The male gathers food for his mate. A technique for feeding, which ensures equality of distribution, is practiced by the young. The chicks form a circle. The one nearest the dim light coming from the tunnel entrance is the first to be fed. This is the first position at which the parent will arrive when bringing food. The chicks then rotate to bring the next chick into the feeding position. This is in sharp contrast to feeding in some other species in which nest mates may be cannibalized during food shortage (snowy owl). In some species, the eldest chick is the only one fed (brown pelican).

---

[29] There is a charming story about kingfishers in Greek mythology. Alcyone, daughter of Aeolus, god of the wind, married Ceyx. They playfully referred to themselves as Zeus and Hera, chief of all gods and his wife. Zeus became enraged at their impudence and sent down a thunderbolt destroying Ceyx's sailing vessel, drowning him. Alcyone threw herself into the sea, committing suicide by drowning. The gods in their compassion changed Alcyone and Ceyx into halcyon birds (kingfishers). Halcyon birds built their nest on the open ocean. (Not true, but what do you expect from mythology?) Aeolus, Alcyone's daddy, calmed the winds during the breeding season for halcyon birds. This is the two weeks preceding the winter solstice. This is the source of the expression "halcyon days," meaning tranquil, pleasant times. This gave rise to the species name of the belted kingfisher, *alcyon*, and occurs in the family name of the tree kingfishers, Halcyonidae. The pygmy kingfisher (*Ceyx pictus*) is only 4.7 in. (12 cm) long and weighs 0.3-0.6 oz. (9-16 g).

Fledglings leave the burrow in about twenty-eight days from hatching. They remain with their parents for another three weeks. Parents teach their young to fish. The parent will drop a fish into shallow water and encourage the young to dive for it.

A few other kingfishers are of special interest. The sacred kingfisher (*Todiramphus sanctus*) very aggressively guards its territory. Animals which approach too close to its nest may have their eyes stabbed out. The laughing kookaburra (*Dacelo novaeguineae*) has a loud laughing call often used in the soundtrack of jungle movies. The jungle in the movie is supposed to be African or South American, but the laughing kookaburra lives only in Australia. Both the blue-winged (*D. leachii*) and laughing kookaburra are snake hunters. The kookaburra grasps the snake behind the head, beats the head against a rock, and swallows the snake head first. Another technique the kookaburra uses to "tenderize" the snake is to drop it from a great height. A large portion of the snake may hang out

Laughing kookaburra
Australia

of the bird's mouth. The kookaburra digests the snake a bit at a time until the entire snake has been swallowed. Most kingfishers are solitary except during the breeding season. The blue-winged kookaburra parents live in a family group with the current brood and some of last year's brood. The year-old birds help in raising the new brood. Interestingly, it is mainly the uncles that help in raising the chicks and not the aunts!

Pigeons and doves constitute the bird family Columbidae. Commonly, the word pigeon is used for larger members of the family and doves for those smaller. There is no scientific significance associated with the two terms. The species usually referred to simply as pigeon is the feral rock pigeon commonly found in great numbers in many cities around the world. There are approximately 280 species of pigeons in the world.

Pigeons nest in trees, on ledges, under bridges, or on the ground. The nest is crudely constructed of sticks and detritus. One or two eggs are laid, and both parents care for the young, called squabs. Pigeons have a unique way of drinking among birds. They drink with their head down, the liquid being propelled by peristalsis of the esophagus. Other birds raise their heads in the air and let gravity guide the liquid into their throat, an action called tipping up. Another exception is parrots which drink by lapping water like a dog or cat. Pigeons have bifocal eyes. The top and bottom of their eyes focus differently.

Quite unusual among birds, pigeons, male emperor penguins, and flamingos[30] produce "crop milk." In pigeons and flamingos, both male and female feed crop milk to infant birds. This "milk" originates from fluid-filled cells which slough off from the lining of the crop. The crop is a pouch at the front of the throat and is used by grain-eating birds to store food. Pigeons eat grain along with fruit and plants. The milk is produced in both pigeons and flamingos under the influence of prolactin, a hormone also present in lactating mammals. By definition, mammals have mammary glands used to nurse their young. What are these birds doing with the hormone, prolactin, in their bodies? Do they think they are mammals? Perhaps this is another interesting case of evolutionary convergence (an adaptation independently developed by unrelated species).

Most birds possess a crop, which mainly functions to store food. When a bird finds a plentiful source of grain, it eats its fill quickly so as not to become a meal for a predator. By using its crop, it can take advantage of the food source by eating more than it can hold in its stomach.

The extinct dodo, which lived on Mauritius Island east of Madagascar, is a relative of the pigeon. The passenger pigeon of North America was once the most numerous of all birds in the world. One breeding colony was estimated to have 100 million birds in 1871. The species became extinct in 1914.

Pigeons are famous for their homing instinct. They use various clues to find their way. It has been thought that they can sense and navigate by the earth's magnetic field. They can also be guided by landmarks. Like many other birds and insects, pigeons can discern the polarization of light from the sky and use it in navigation. It is well known that pigeons have been used in time of war to deliver messages. During WWI, a pigeon named Cher Ami ("dear friend" in French) delivered a message, which saved the lives of 194 American soldiers. During WWI at the battle of the Argonne in October

---

[30] Flamingos feed on brine shrimp which they pass through a strainer in their beaks to separate food from mud. The flamingos feed with their head under water, pointed backward, and held upside down. The flamingo crop milk is produced throughout the upper digestive tract. Both male and female parents "nurse" their chicks with this milk until they are able to filter feed. Flamingo milk is richer in fat and protein than is pigeon milk. Their milk also contains red and white blood cells as well as a red pigment stored in the liver.

1918, the "Lost Battalion" of five hundred American soldiers was trapped behind enemy lines. They were receiving friendly fire from their own troops, who didn't know their location. By the second day, only about two hundred still survived. They sent a pigeon with the message, "Many wounded. We cannot evacuate." The German enemy shot this pigeon out of the air. They sent a second pigeon with the message, "Men are suffering. Can support be sent?" This pigeon too was killed. Cher Ami, the only pigeon left, was dispatched with the message, "We are along the road parallel to 276.4. Our own artillery is dropping a barrage directly on us. For heaven's sake, stop it!" The men watched as Cher Ami was shot down, but then he was airborne again. He arrived at headquarters 25 mi. away in twenty-five minutes. Cher Ami was blinded in one eye, shot through the breast, and arrived with one leg hanging by a tendon. The intrepid pigeon saved the lives of 194 American soldiers. For Cher's bravery, it was awarded the *Croix de Guerre*. Medics worked to save the bird's life, and Cher Ami survived for a year before dying of his wounds. They fashioned a wooden leg for him. General John Pershing personally attended the bird's departure for America. Cher Ami was mounted by a taxidermist and can be seen on display with Sergeant Stubby, a heroic war dog, in the National Museum of American History's "Price of Freedom" exhibit.

Chapter 14

Mammals (Land and Sea)

Monotreme derives from the Greek, monos, meaning "one" and trema, meaning "hole." It designates a group of mammals which possesses a cloaca, a single outlet for excreta from the gut, kidneys, and sperm or eggs. There are just two kinds of mammals in this group, the duck-billed platypus[31] and the echidna. There are two varieties of echidnae, the short nosed and the long nosed. The platypus and short-nosed echidna live only in Australia. The long-nosed echidna lives only in New Guinea. Monotremes all lay eggs; they are the only mammals to do so. Mammals do not normally have a cloaca. Aside from monotremes, the only mammals having a cloaca are the marsupials. Marsupials are not very familiar animals unless you live in Australia, where there are plenty. There are some in North and South America. In the United States, there is just one marsupial, the Virginia opossum. There are several types of possums in South America (not a spelling error, opossum in the north and possum in the south). Both the platypus and echidna are endothermic (warm-blooded), but both have lower body temperatures than other mammals, 90°F (32°C) for monotremes, versus 99°F (37°C) for other mammals.

---

[31] In 1842, the great American showman, P. T. Barnum, exhibited a stuffed duck-billed platypus in New York and then around the country. He called the animal by its proper scientific name, *Ornithorhynchus anatinus*. A companion exhibit was the "Feejee" mermaid, the upper body of a female orangutan mated to the tail of a large fish. Financially, the tour was quite successful, but both the Feejee hoax and the platypus were condemned as fakes, an embarrassment to Barnum. When first exhibited in Europe by others, mounted specimens of the platypus had also been decried unjustly as hoaxes.

Echidnas are covered with spiny spikes like a hedgehog. They principally eat ants and termites. Three or four weeks after mating, the female echidna lays an egg no larger than a small grape. The egg comes to rest in a small depression on the echidna's belly and sticks to her fur. She carries this egg for about ten days before it hatches. The infant slits the leathery shell open using an "egg tooth" in much the manner of birds and some other reptiles. The baby weighs only half a gram. Such an undeveloped offspring at birth is only known among marsupials.

Short nosed echidna

The baby climbs to an area on the mother's body with special hairs where milk oozes out from mammary glands. The baby drinks from the pool of milk. There is no teat on which to nurse. The echidna has a groove in which she carries the newborn for a time, another great similarity to a marsupial. She later digs a hole in the ground and deposits the baby. She goes about feeding herself and returning to nurse her baby until the infant is able to forage for its own insects.

The mainly aquatic platypus builds a deep, complex tunnel in a riverbank where babies can stay. When the mother platypus leaves young or eggs, she seals the tunnel entrance with mud. Since she is swimming most of the time, she could not keep babies in a pouch even if she had one. The duck-billed platypus eats small creatures it finds on the river bottom, freshwater shrimp, insect larvae, or small mollusks. The rubbery bill of the platypus is equipped with delicate sensors which respond not only to touch, but also to the electrical fields generated by the nerve impulses of its prey. Such electrical sensors are common in fish, but unknown in mammals. Neither the echidna nor the platypus has teeth. They both grind their food against a hard surface. This surface is in the roof of the echidna's mouth or in the bill of the platypus.

Wildebeests are ruminants and survive by eating grass. They also must have water and can only survive for one or two days without drinking. They spend the wet season on the plains of the Serengeti in Tanzania. The wet

Wildebeest in African grassland

season comes to an end in May, forcing the wildebeests to move northward to greener pastures. They cross the Mara River on the way to the Masai Mara plains in southern Kenya. They remain until rains again start to fall near the year's end. Strangely, just when grass is about to show renewed growth; they head back toward the Serengeti. Submerged crocodiles lurk, only their eyes and nostrils showing, along the banks of the Mara River. Crocodiles will take hundreds of wildebeests, but thousands will cross. Why do they leave their green pastures in Kenya? Scientists believe it is because of the dietary requirement for phosphorus. Phosphorus is available in the volcanic soil of the Serengeti, but not in the Masai Mara. Wildebeests are more or less constantly on the move, covering 500-1,000 mi. (8001600 km) every year.

The female wildebeest calves in the midst of this migrating herd, as opposed to seeking seclusion like most antelopes. Though she gives birth while lying on her side, she arises immediately afterward, severing the umbilical cord. (Some authorities state that she delivers standing up and show photographs to support this contention.) About 80% of females calve within the same two- to three-week period. Some females may employ delayed implantation of the embryo in order to accurately time calving. This helps to ensure survival of enough calves in spite of some being lost or devoured by lions, hyenas, or other predators. It may also decrease predation. The individual calf has a better chance of survival when there are more calves from which to choose. Within a few minutes, the young wildebeest is following Mom on this strenuous journey. Within a few days, the young can run fast enough to keep pace with the herd. Such completely developed young are said to be precocial.

Kangaroos are marsupials, a suborder of mammals, which give birth to offspring at a very early stage of development. Most marsupials have a pouch (marsupium) for carrying their young. There is a wide variety of species (some 250) of various sizes and habits. Examples are large red and gray kangaroos 200 lb. (90 kg), wallabies 25 lb. (11 kg), rat kangaroos, koalas, tasmanian devils, wombats, tree kangaroos, opossums, and gliders. The oldest fossils of marsupials are found in North America, where they first evolved. They seem to have been displaced from North America by placental mammals. Placental mammals give birth to fully developed young since the placenta provides adequate nourishment to support a full-term pregnancy. Some marsupials outran the mammals, going to South America and thence to Australia by land bridges which still existed at that time. At least this is the prevailing theory. Marsupials are also found in New Guinea and New Zealand. Marsupials may be better able to survive than do other mammals the frequent droughts that plague Australia. The mother marsupial has much less energy and food resources invested in her offspring than a mammal which has carried the baby until fully developed. During drought, she can abandon the offspring, sacrificing only a smaller investment. Kangaroos are known to cease mating during droughts but immediately to mate once the drought breaks. The red kangaroo drinks water only during times of severe drought. During droughts as many as 70% of kangaroos perish. The kangaroo digs holes in the ground in order to find water. Wild pigeons and emus drink from these holes.

The familiar red kangaroo will serve as an example of the kangaroo tribe. The red kangaroo is the largest of the surviving marsupials. You may have encountered references to this kangaroo exploited in boxing matches with professional boxers. This form of entertainment was outlawed in the early twentieth century. This very common animal in Australia feeds on grass and, as mentioned earlier, is a foregut digester which does not produce methane as do sheep and cattle. Though not considered to be a ruminant, the kangaroo does regurgitate some of its food and rechews it. Its life is limited to the time it takes to wear out four sets of molars. Unlike the elephant all of the kangaroo's molars are erupted in the adult. As the foremost molars wear out they are discarded and are replaced as the rest of the molars advance. Once the kangaroo has no more molars, it starves to death.

Kangaroos are hunted for meat and hides. The meat may be consumed by humans, but is commonly sold as dog food. Red kangaroos can hop along at a clip of 35 mi./hr. for short distances. The kangaroo has five toes; the fourth, being the largest, is used as a platform from which to hop. The first toe is almost invisible. It is vestigial and of no use. Toes 2 and 3 are fused for nearly their whole length (syndactyl) and are used for grooming.

Kangaroo frightened
by photographer retreats to water

Kangaroos are sometimes killed as pests because they compete for grazing land. They are also hunted by wild dogs (dingoes). When cornered by dogs against a fence[32], the kangaroo panics and is killed, though it could

---

[32]   Kangaroos are more plentiful on the south side of the great Dingo Fence that runs east and west for 1,500 mi. (2,400 km) through Queensland and South Australia. This fence is the longest in the world. It was originally built in 1884 in an unsuccessful effort to contain the spread of rabbits. It was modified in 1914 to make it dog- proof to stop dingoes from crossing. The fence is 5.9 ft. (1.8 m) high and can be cleared by an adult kangaroo. A staff of twenty-three employees maintains the fence. Both rabbits and dogs (which gave rise to the wild dingoes) were introduced to Australia by humans.

easily have leaped over the fence. This is thought by some to illustrate low intelligence. Similar behavior is exhibited by the supposedly very intelligent porpoise. When a purse net closes on a porpoise, it drowns though it could easily have jumped over the cork floats. One strategy employed by red or gray kangaroos when pursued by dingoes is to head for a body of water, if there is one nearby. The kangaroo will move into waist-deep water, and when a dog comes within reach, the kangaroo will seize the dog with its front arms and force it underwater to drown the predator. In the accompanying illustration, a kangaroo became frightened as the photographer tried to approach closer. The kangaroo fled into the surf just as described above as if being pursued by a dingo.

Kangaroos pant to cool themselves. They also lick their arms, which are supplied with a dense bed of capillaries close to the surface. The evaporation of their saliva has a cooling effect. The kangaroo's tail acts as a storage depot for fat when food is plentiful. The large tail moves up and down like a pump handle as the kangaroo hops along. Its hopping mode of running is more energetically efficient than motion on all fours. The motion of the tail jostles the viscera of the kangaroo which aids in breathing so that less energy is expended in breathing.

The female kangaroo is capable of mating immediately after giving birth. It takes little effort to deliver jellybean-sized young. The fertilized egg develops to the blastomere stage consisting of about one hundred cells and then becomes dormant. This arrest of development is called embryonic diapause or delayed implantation. For further development to occur, the embryo must be implanted in the wall of the womb. While this dormant embryo is present, the female may be caring for two babies previously born, one (called a joey) in her pouch and the other "at heel." It is important that development of the new embryo be arrested because there is no room for another baby in the pouch. A new arrival would be so delicate that it would be crushed by the joey already in the pouch. Delay of implantation also gives the joey in the pouch time to be old enough to leave the pouch. The joey in the pouch is living off mother's milk and so is the joey at heel. The joey at heel does not reenter the pouch, but stands outside to nurse. The two siblings have different milk requirements. The mother has four teats, and the two joeys nurse from different teats, which produce two different kinds of milk. The joey at heel gets milk of a higher fat content. When the demand for milk slakes off a bit, this signals the time for implantation of the blastomere into the wall of the womb and continued development of the next baby. When the new one arrives, Mom will kick the joey out of the pouch. The older joey can now subsist on grass and so is weaned

At birth, an embryonic allantois is first passed. This is a small spherical watery sac about 2 cm (0.78 in.) long and contains the excreta from the embryo. The baby emerges and tears loose from the fetal membrane, using the sharp claws on its front feet. The 0.75 g (0.026 oz.) baby climbs unaided 6-8 in. (15-20 cm) up to its mother's pouch. This takes two or three minutes. The mother's pouch opens upward so that the baby won't be thrown out as the mother hops about[33]. The female kangaroo has the ability to close the pouch tightly and has been known to swim across a water hazard when pursued by predators. Her pouch is closed tightly enough that the joey inside is protected from drowning.

Tasmanian devil

The tasmanian devil is a marsupial which inhabits the island of Tasmania, off the southern coast of Australia. Americans are familiar with the comic character called by this name or affectionately known as Taz. Many Americans display Taz as tattoos somewhere on their person. The feisty Taz is fierce and bellicose. The real tasmanian devil is not that different from the comic one. Its guttural growl is unworldly.

The animal is black with white marks on its chest and hindquarters. The tasmanian devil is the largest carnivorous marsupial and is 30 in. (76 cm) long and weighs 26 lb. (12 kg). The devil has sharp teeth and very powerful jaws. Farmers on mainland Australia believed that the devil was a threat to livestock, though they were only proven to take chickens. They were probably eliminated from Australia by dingoes. In 1941, the government passed laws protecting the tasmanian devil. Their numbers have increased under this program.

Tasmanian devils are strict carnivores, eating snakes, birds, fish, and insects. They are also scavengers and will, in groups, eat a large carcass whenever available. They are nocturnal and find food through keen sense of smell. They use their long whiskers like a cat to find prey. When they eat, they consume every part of the prey or carrion including hair, skin, and bones. They crush the bones with their sharp teeth and powerful jaws.

---

[33]   The koala bear (not really a bear) has a pouch opening downward. The koala lives in trees, moves slowly, and sleeps fifteen hours a day so that the baby is not in danger of falling.

In the decade of the 1990s, it was discovered that a disease was threatening the tasmanian devil. The animal is subject to a rare, contagious facial cancer. This cancer eventually interferes with the animal's ability to eat and results in fatal starvation. Animal health workers are combating this threat by sequestering healthy tasmanian devils and are using captive breeding to prevent their extinction.

Pandas have what functions as an opposable thumb with which it can grasp its food. This pseudodigit is a special adaptation of a wrist bone. Male pandas weigh 250 lb. (110 kg); females about 180 lb. (80 kg). The panda eats bamboo shoots and leaves and practically nothing else. In spite of being an herbivore, the panda is usually classified in the order Carnivora,

Giant panda holding bamboo

and family Ursidae (bears)[34]. Panda classification is still being studied. Some authorities believe the panda is more closely related to the raccoon[35]. DNA evidence seems to support the classification as a bear. The pupils in a panda's eyes are vertical slits like those of a cat. The Chinese people call the panda *da xiong mao*, meaning "giant bear cat." The seven other species of bear have round pupils.

Zoologists call the panda a specialist for having such a restricted diet. Pandas eat almost exclusively just two species of bamboo, arrow and umbrella bamboo. Bamboo plants are subject to periodic die-offs after flowering, and all of the bamboo plants in an area come into flower simultaneously. Sometimes this leaves the pandas with nothing to eat, and they starve. Formerly, pandas would migrate to another area to find patches of bamboo still flourishing. Now the panda's range is fragmented, and this is not possible. Pandas live in isolated colonies which do not interbreed. This is a classic case of the danger of dietary specialization, subsisting on just one food source.

It is thought that the panda evolved from flesh-eating predecessors. Meat is easy to digest with the enzymes possessed by carnivores. For this reason, carnivores have shorter guts than herbivores do. The panda, carnivore turned

---

[34] In astronomy, Ursa Major is the Great Bear constellation.

[35] The smaller red panda is considered to be in the raccoon family. Like the giant panda, the red panda has a specialized wrist bone which acts as a thumb for grasping bamboo which is its only food. The red panda has rings on its tail like a raccoon and dark patches over its eyes like the giant panda and the raccoon.

herbivore, kept its short gut and rapid digesta transit time. This small gut and short dwell time of digesta prevents the panda from making much use of bacterial fermentation, which saved the day for other large herbivores. It's a wonder that the panda has survived to this point. Pandas ignore other kinds of plants which are available, though they could digest them more easily than bamboo. The panda could eat meat but rarely does so. The panda's useful range has been severely curtailed by human intrusion so that the panda is endangered. Because the panda is a national symbol, the Chinese government is trying, by setting aside protected areas and by a captive breeding program, to save the panda. Let us hope it is not too late.

Armadillos are mammals with a scaly protective shell. A number of bands encircle the body of armadillos. The underside of an armadillo is not protected with scales and has meager distribution of hair. The nine-banded armadillo's scientific name is *Dasypus novemcinctus*[36]. Typically, it has nine bands, but may have as few as seven or as many as eleven bands. There are about twenty species of armadillos living in Mexico and South America, but the nine-banded is the only one living in the United States. The discussion here will deal with the nine-banded armadillo.

People living in Southern United States most frequently become aware of armadillos in the form of roadkill. The armadillo's response to being startled is to leap vertically 3 or 4 ft. (0.3 or 1.2 m) in the air. When a car startles the animal, it ends up colliding with the undercarriage of the vehicle. This might scare a predator long enough for the armadillo to hurry away, but cars show no fear[37].

Nine-banded armadillo in grass

Armadillos dine principally on insects and their larvae. Armadillos are very well equipped for digging. Armadillos become more active in the evening and at night. They often live in burrows having several chambers. They willingly share some compartments with burrowing

---

[36]   Dasypus derives from the Greek word for rabbit. The novem part of the species name is reminiscent of the month November. Novem is Latin for nine as in *nine*-banded. November was the ninth month in the Roman calendar. The cinctus part of the name recalls the word "cincture" or "band," as in nine-*banded*.

[37]   The armadillo's method of dealing with automobiles reminds one of the skunk's technique for handling this problem. As a car approaches, the skunk holds his ground and sprays and then dies.

owls, rabbits, or rattlesnakes. Hunters, who have reached into such burrows, hoping to haul out an armadillo by its tail, may instead withdraw their hand quickly after suffering a rattlesnake bite. Remember, rattlesnakes can see in total darkness.

The armadillo has an interesting way of crossing bodies of water. When the armadillo encounters a shallow ditch, it gulps air and fills its esophagus. With just a little air, the armadillo can sink to the bottom and walk across the ditch. With the air supply, the armadillo can hold its breath for a good five minutes. If the animal needs to cross a more substantial water hazard, it gulps much more air into its stomach and intestines and floats across.

Persons living in eastern Texas and in Louisiana would sometimes eat armadillos[38]. They also manufactured souvenirs from armadillo shells and handled the animals for the sport of "armadillo races." When persons involved in these activities began coming down with leprosy (Hansen's disease[39]), medical workers discovered that leprosy was carried by wild armadillos. Eating armadillos is not recommended.

Researchers studying the disease of leprosy had faced a serious problem. You couldn't induce leprosy in a human being; that's not ethical, and besides, no one would sign a consent form. The usual strategy is to give the disease to some lab animal—such as a dog, rat, or guinea pig—or to grow the bacterium in vitro (i.e., in a petri dish). The trouble was that leprosy wouldn't grow in any of these animals; neither would it grow in vitro. Leprosy tends to show up first on the ears, fingers, or noses of human victims, areas where the temperature is slightly lower. Armadillos have a low body temperature, 93°F (34°C), and easily become infected with leprosy. In the nine-banded armadillo, leprosy attacks the brain, spinal cord, and lungs. These organs are normally spared in human lepers. Louisiana was the site of the Carville institute where lepers were treated and important research was conducted from 1894 to 1998. Though there were many name and administrative changes, it was usually called Carville. The institute was moved to Baton Rouge, Louisiana, in 1998. The original site became a museum.

The nine-banded armadillo has a unique reproductive strategy. The female always gives birth to precocial identical quadruplets, which shared a common placenta. Only nine-banded armadillos show this behavior and not armadillos

---

[38] German immigrants in the United States called the nine-banded armadillo Panzerschwein, "armored pig." During the Great Depression of the 1930s, people called the armadillo a "Hoover hog" because citizens resorted to eating armadillos due to hard times. This was in contrast to Hoover's promise of "a chicken in every pot, a car in every garage."

[39] G. H. Armauer Hansen, a Norwegian physician, established in 1873 that a bacterium (now called *Mycobacterium leprae*) was the cause of leprosy. This was the very first instance in which a bacterium was established to be the cause of human disease.

in general. Production of quadruplets has served the armadillo well. This provides scientists with the opportunity to determine characteristics which are inherited or modified by experience. One difficulty with this approach is that nine-banded armadillos are reluctant to mate in captivity.

The capybara is the world's largest rodent, weighing up to 165 lb. (75 kg). The capybara has the appearance of a giant guinea pig with a large head, blunt nose, and only a trace of a tail. Capybaras are semiaquatic and have partially webbed feet. They eat grass and live near the water. When alarmed, they bark a warning and run for the nearest body of water, seldom far away. Capybara in the Guarani Indian language means "master of the grasses." The capybara eats some of its own feces, in common with other rodents and in common with rabbits (lagomorphs, not rodents.) This habit may increase the population of bacteria needed to digest grass. Mammals do not have enzymes of their own to digest grass. They may also gain needed vitamins from feces produced by bacteria at night in the cecum but not produced in the daytime during active feeding. (The word may is meant to imply author's conjecture.)

Capybara with young at water's edge

In Argentina and Venezuela, capybaras are raised side by side with cattle. Once a year they are harvested and used for human food. Centuries ago, one of the popes (which one is uncertain) decreed the capybara to be classified as fish so that Catholics were permitted to eat capybara on Fridays and during lent. The legal basis for such a seemingly strange conclusion relates to the *Summa Theologica* of Saint Thomas Aquinas. This treatise argues for classifying animals by living habits as well as by anatomy.

The beaver is the world's second largest rodent. The beaver is champion at building dams. One dam in Northern Alberta is 2,790 ft. (850 m) long. The entire beaver family participates in building the dam. They may be aided by neighboring families in dam construction. Beavers can fell trees 9.8 in. (25 cm) in diameter. They also construct canals to float building materials too heavy to be carried to the site of the dam. Beavers use mud and rocks to make the base of the dam. They excavate mud from the streambed and carry it against their chest with their front paws. Beavers are capable of walking erect on land while carrying materials in their arms.

The dams often improve the environment for wildlife by providing wetlands for many species. Wetlands have suffered great losses because people like to live near water and build on wetlands in spite of legislation to discourage this practice. Officials may close one eye to this practice. The other eye opens to peek at property taxes they can collect. By slowing the flow of water past the dam, silt settles to the bottom, thus clarifying the water and improving it as habitat.

Beavers, like rats, are gnawing animals. Their incisors grow throughout life and are kept at the proper length by an automatic sharpening while they chew. No animal chews more wood than the beaver does. The incisors are kept sharp

Beaver on dam

in the same manner as described above for rats (chapter 3). In the space behind the incisors, there is a flap of tissue on each side of the mouth that can be made to meet in the middle, closing off the throat. This prevents wood chips from entering the beaver's throat. When underwater, this closure prevents water from entering the beaver's throat. The beaver gnaws through wood below water as well as in the air. This protective flap of tissue is also found in the rat.

A third inner eyelid, the nictitating membrane, closes over the eye when the beaver is underwater or chipping away at trees. This protects the beaver's eyes from wood chips. The beaver's ears and nostrils can be tightly closed while underwater.

Beavers are stimulated to build dams by prolonged exposure to the sound of running water. Beavers have been observed to pile building materials close to a loudspeaker broadcasting the sound of running water, but only after continued exposure to the sound. Beavers continue to repair the dam and build it higher as long as the sound persists. At high water times, beavers will permit spillways to flow freely. The beaver's tendency to fell trees appears to be instinctual since they persist in felling trees even if they have no place to build a dam.

Beavers continue growing throughout life. Beavers eat tree bark, twigs, and the roots of aquatic plants. Beavers are 3-4 ft. long (91-120 cm) including the tail. They weigh from 40 to 95 lb. (18-43 kg). Females are as large as or larger than males of the same age. This is unusual for mammals. Beavers have a broad scaly tail, which they use as a rudder while swimming. When alarmed, the beaver slaps the water with his tail. The report sounded travels for great distances both above and below water. The beaver then dives to hide from danger. All the beavers in the area do the same, not emerging from the water for some time. Beavers can swim for half a mile underwater and can remain submerged for fifteen minutes.

Beaver fur has been of great commercial value especially during the nineteenth century. Much of the impetus for exploration of North America was due to the lure of the fur trade. Beaver furs were used for clothing. Manufacture of fur-felt hats from beaver fur mixed with other furs involved the use of mercury compounds and led to mercury poisoning in workers so employed[40]. A disease called the Danbury shakes afflicted fur-felt hatmakers in Connecticut. The glands and testicles of beavers have been used in perfumery and as an analgesic. The pain-killing effect has been attributed to the willow trees in the beaver's diet. Indians had used willow bark infusions in pain management. The genus name for the willow is Salix, related to salicylic acid, used in the production of aspirin.

The main advantage that the dams provide for beavers is protection from predators such as coyotes, wolves, and bears. In the middle of the lake that forms above the dam, the beaver builds his home called a lodge. The lodge is made of branches, plants, and mud. Multiple entrances to the lodge are beneath the water so that the beavers can come and go unseen.

Young beavers live for two years with their parents. When a newborn arrives, the two-year-olds are forced to leave the lodge to seek their own territory. Beavers rarely fight except when the young are driven away. Disbursal of the young aids in preservation of the species. In many species, this tactic prevents extreme competition for food. It also leads to a spread in the range of the species.

By the late 1800s in America, there were very few beavers left due to long and intense trapping for their fur. Since the start of the twentieth century, wildlife in general has benefited greatly from the conservation movement and the establishment of national and state parks.

The bishop of Quebec brought up the question of whether or not beaver might be classified as fish. Beaver had been on the menu of the indigenous people before the arrival of Europeans. The Roman Catholic Church officially granted a fish designation for the beaver.

The moose is the largest member of the deer family. The plural of moose is moose. Males are 6-7 ft. (1.8-2.1 m) tall at the shoulder and weigh 850-1,580 lb. (380-720 kg). Females are smaller, 600-800 lb. (270-360 kg). Moose are second in size only to the bison among land animals of America.

The range of moose in America is greater now than at any time in recorded history, largely due to conservation efforts. Its range includes almost all of Canada, Alaska, the New England states, Northern Rockies, Northeast Minnesota, Upper Peninsula of Michigan, and pockets in Utah

---

[40]   Some say that mercury poisoning afflicted the Mad Hatter in the tale *Alice's Adventures in Wonderland*. Mercury, like other heavy metals such as cadmium and lead, attacks the nervous system.

and Colorado. In the Old World, moose range across Great Britain, northern continental Europe, and northern Asia.

Bull moose in velvet standing in water

In Europe, the animal is called elk. The name moose comes from the Algonquin *moz*, meaning "twig eater." Moose eat twigs of birch, aspen, and especially willow. They like to graze on flowers and other plants and to browse on twigs of willow. Moose eat the catkins of willows, which make their appearance before willow leaves in early spring. They seek out licks to supply themselves with minerals, particularly sodium needed to grow antlers. Moose feed on water lilies and other water plants, which contain sodium. Moose also are known to lick the salt (sodium chloride) off roads in winter months. They have a unique ability among deer in being able to eat aquatic plants underwater without raising their heads. All deer can pluck plants from the water but must raise their heads in order to swallow. Exactly how this works is a bit of a mystery. The moose's large soft nose is a very complex organ and is, in fact, a distinctive feature of the moose.

The moose's diet is complex. A moose cannot be maintained on just some artificial moose pellets manufactured with the moose's nutritional requirements in mind. Partly for this reason, moose have never been domesticated. Attempts to domesticate moose have met with, at best, mixed results. In the late fall, after the rut, or breeding season, both male and female moose eat as much as possible to prepare for winter. Males drop their antlers to lighten their load by 50 or more pounds (23 kg), thereby saving energy. Females that may be pregnant need extra food to supply the growing fetus.

During the winter, food is scarce. They live in cold climates but do not hibernate, and even though they spend much time eating twigs, they cannot get enough food to maintain good body weight. Both sexes waste away and become weak during the winter, a time when they struggle just to stay alive. This winter slump in health is another strike against domestication of moose. Moose hair is hollow, which makes it better for heat insulation. Many that become too emaciated, especially the old and the very young, are food for wolves.

In the spring, new plant growth again provides nutrition for the depleted bodies of the moose. Both males and females need to regain lost weight. Moose of both sexes need an abundance of minerals. Females need to nourish their unborn calf or calves and to produce milk. Females have lost some bone mass, which has to be replaced. Twins are not unusual, in which event Mom must eat for three. Males need to regain bulk and strength and to grow new antlers, crucially needed for next fall's rut. Female moose do not have antlers. A good rack of antlers helps a male gain the opportunity to mate and pass his genes on to progeny. Presumably, a good rack of antlers is a sign of a healthy male moose. Females are choosy about their mate, and part of their inspection involves licking the male's antlers. Antlers are grown each year and reach maximum size at about ten years of age. An area of forest which has been devastated by fire is full of minerals derived from ashes of trees and shrubs. The new plant growth is laden with these minerals and is the favorite forage for moose. Forest fires, logging, and floods, disastrous as they may be, are good for moose. They result in a bonanza in nutrition for the moose.

Antlers may grow at the rate of 1 in. (2.5 cm) per day. Fully developed, the antlers may be 5 ft. wide (150 cm) or more. The blood supply to the growing antlers is provided by a densely vascular coating of "velvet." The velvet is alive and sensitive. The moose is careful not to damage the tender velvet as this would be painful and could lead to abnormal growth. By August, growth of antlers has reached its zenith, and the drying velvet is hanging in tatters. The male moose rubs its antlers against shrubs and trees to rid itself of the drying velvet.

The mating season, called rut, occurs in mid-September through November. In preparation for rut, males spar with one another. These are friendly bouts with males, which may be unevenly matched. No one gets hurt, and there is neck twisting. When serious competition for females starts, fights usually occur only with evenly matched contenders. If the match is uneven, the lesser male turns away and leaves and is not pursued. Serious fights involve pushing with brute force and not with neck twisting. Serious injuries are not common but certainly do occur. Puncture wounds from sharp horns are likely as are tears around the head, neck, and rump. A successful male may mate with several females, one at a time. The females may be accompanied by a calf born in the preceding spring, but the male pays no heed to the calf. The male may sample the urine of the female, raising his head and curling back his upper lip in the flehmen gesture. Urine flows through a small slit in his palate onto Jacobson's organ (vomeronasal organ) in much the same way as described above for the giraffe. The male is able to tell from his analysis whether the female is ready to mate. The male urinates on the ground, stomps in it with his front hooves, making a muddy spot and splashing it on his head and bell[41]. The bell is a fleshy fur-covered growth which dangles from his throat. He marks shrubbery with this bell in order to advertise his presence to cows. He wallows in the wet urine spot, as does his

---

[41]   The purpose of this bell has been a mystery. The male has a bell and from it a rope. The rope may freeze during very cold winters and fall off. The female has only a rope.

female consort. The urine contains pheromones which excite the female and help bring her into estrus.

The female requires about 230-240 days of gestation before giving birth. The female seeks seclusion to give birth. She eats the placenta so that predators will not be attracted. The mother suckles her young, lying down, a behavior common only to reindeer among members of the deer family. Moose milk is rich in milk solids (20-24% by volume). Horse and camel milk are more dilute (8-12% by volume for the horse). The mother guards her calf diligently. If you are in moose country in spring or summer, be careful not to approach a mother and her calf. Be especially careful never to get between the mother and her calf; she will attack you.

As the birth of a new calf approaches, the mother chases away any yearling calf. The calf does not leave willingly and tries to rejoin its mother. It has been totally dependent on Mother to protect it from wolves and bears. She again rejects the calf, which follows her at a distance for a few days before departing. This is a vulnerable time for calves as they are at loose ends and may fall prey to wolf or bear.

The moose, a ruminant, chews its cud and makes maximum use of the food it eats. Like all ruminants, the moose has only lower incisors. In place of upper incisors, the moose has a hard plate. Compared to a cow, the moose has a small rumen. If only poor fibrous food is available, this may be a serious problem for the moose. Because food moves slowly through the system of a ruminant, the moose may not be able to eat enough low-quality food to survive, and the moose will starve to death. A moose seeks out and chooses the most nutritious food available. Moose are selective about the terrain, finding the most fertile land available for forage. Young shoots are best and contain the least toxins. The foliage at the top of a tree is less toxic than that lower on the tree. A moose prefers to eat foliage from the top of a tree accidentally felled in a storm. Some of the plants eaten by moose are toxic, for example, willow bark and twigs. Willow happens to be a favorite with moose. Moose have a very large liver to metabolize toxins to forms less toxic. Some natural toxins may accumulate in the liver. In Finland, it is illegal to eat the liver or kidneys of moose more than one year old because of high cadmium content. Cadmium, a heavy metal, is elemental and cannot be metabolized to a less-toxic form.

Moose are a prey animal for wolves. In Sweden, wolves would prefer to attack deer, caribou, or elk rather than moose. Moose can deliver bone-crushing kicks with both front and rear legs. Autopsied wolves sometimes show the results in the form of healed skull fractures and broken ribs. Wolves usually hunt in packs and try to wound the moose and then follow it until it loses enough blood to be less dangerous. The moose has a keen sense of smell and very sharp hearing. The moose likes terrain with many obstacles 2 or 3 ft. (0.6 or 0.9 m) high. The long-legged moose can negotiate such obstacles without breaking stride. A moose can run 35 mi./hr (22 km/hr). Wolves or bears have to leap over such obstacles at considerable expenditure of energy. The wolves could maneuver around the object, which would slow them down considerably. In this way, the moose would probably escape. Snow, if not too deep, say 2 ft. (60 cm), favors the moose. Three feet of snow would make running difficult for the moose. A hard crust on deep snow would be very difficult for the moose, but easy for the wolf which can be supported by the crust. Such a crust occurs late in winter when the sun melts some surface snow which then freezes again. Such a crust is called a sun crust. Moose under such conditions dig out extended networks of trenches called moose yards. Bears also hunt moose, but would prefer to steal a wolf kill, rather than kill the moose themselves.

The main predator of moose is man. Rock drawings and cave paintings in Europe show that moose were hunted from the time of the Stone Age. In northern Scandinavia, one can still find the remnants of moose pits used to trap moose. The pits were lined with steeply inclined planks to prevent the moose from climbing out of the pit. There are also the remains of wooden fences used to direct the moose toward the pit. This method of hunting was very effective and was used up until the start of the nineteenth century when it became illegal.

Moose do not fare well in captivity. It is very difficult to manage their nutrition. They cannot survive on a simple diet. Moose are very susceptible to parasites carried by livestock or by white-tailed deer. The white-tailed deer carries three parasites particularly destructive to moose: meningeal worm, winter tick, and giant liver fluke. The meningeal worm, a nematode or roundworm, settles in the white-tailed deer's meninges. The meninges consist of three membranes enclosing the brain and spinal cord. The meningeal larvae are picked up by a snail and finally by the moose. This does not cause much distress for the deer. However, when a moose becomes infected, the worm does not stay in the meninges but invades the brain tissue with disastrous results. Winter ticks multiply and may denude the moose of its winter fur. These naked moose are called "ghost moose". Moose have not fared well in zoos. Artificial pellet-type food is not adequate. In the Milwaukee County Zoo, moose do well on a variety of fresh browse (willow, maple, mulberry) provided by tree surgeons.

Moose have qualities which would make them very useful as domesticated animals. They give good milk, provide meat, are very good at transporting loads, can pull heavily laden sleighs, and are excellent mounts for human riders. They also get on well with humans and are extremely loyal to their keepers. It is difficult to feed them properly, and they are susceptible to diseases from livestock. They cannot be used as beasts of burden or mounts during the winter and spring because they are too run-down and weak during these times.

Horses, not accustomed to moose, are terrified by them. This relationship has at times been exploited. Czar Ivan the Terrible sought to conquer Siberia. The Siberians rode astride moose; the Russians rode horses. When the two animals met, the horses bolted and ran away. The Russian general, Timofeitsch, hunted down the moose riders and had them publicly flayed, mutilated, and impaled. The fact that he went to all this trouble shows that he considered the moose important. Moose were banned in one Estonian city, Dorpat, because of the commotion they caused when horses scattered uncontrollably at the site of them.

Some people have made the mistake of befriending an orphan moose. A moose less than a week old will readily follow anything that moves, treating that being as its mother. This is a rather common behavior first studied in depth by Konrad Lorenz, Nobel laureate, who helped found the study of ethology (the study of animal behavior). Baby ducklings followed Lorenz as they would a mother duck. Baby ducklings will follow a pig in the same way. When grown, they tried to mate with him. This attempt at mating also occurs with people who try to raise a baby moose. A pet moose will not tolerate being left alone in a stable. Its incessant cries can disturb the dead. Inevitably, it will be let in to the house. The little moose, the size of a large dog, is cute, frisky, and amusing. The moose curls up by the fireplace, just like a dog does. It is curious just like a cat. Soon it is tipping over vases, eating the flowers, and getting into everything. In five months, it will weigh 300 lb. (140 kg). When forced to stay outside, the crying starts at the back door and lasts through the night for perhaps three nights. The moose is very loyal to its keeper and will recognize him immediately after a long separation. If the moose is female and is used to provide milk, the moose will be fiercely loyal to its keeper, even killing its own offspring and lavishing affection on its keeper.

Camels are ungulates, that is, hoofed animals. They have just two toes on each foot. There are two species of camel, the dromedary having one hump and the bactrian having two humps. The dromedary lives in the Middle East and Northern Africa, where all dromedaries are domestic animals. There is a feral population of dromedaries in Australia where the camel was introduced as an exotic species. The bactrian lives in the deserts of central Asia, where

Two dromedary camels in desert

temperatures are sometimes below freezing. Both types of camels are domesticated, though there is a small population of wild bactrian camels in Mongolia and China. This discussion will center on the dromedary.

Dromedary camels were introduced into Australia in the nineteenth and early twentieth centuries as transport animals. A feral population grew to perhaps seven hundred thousand. This population has been increasing by 11% per year. In South Australia, culling has been adopted by means of aerial marksmen. Camels are competing with sheep for grazing land.

A number of dromedary and bactrian camels were imported from Turkey to form the U.S. Camel Corps experiment. The camels were to be used as draft animals in mines. They escaped or were released after the project was abandoned. By the 1900s, the camels had died out, though a sighting of a camel was reported in Los Padres National Forest in 1972.

Persians used camels in combat against Lydia. The camels scare off horses when at close range. Horses detest the odor of camels.

Dromedary camels were domesticated as early as five thousand years ago. They have been called "ships of the desert" and to this day are used for transport across the Sahara. They can carry loads of 200 lb. (90 kg) and travel 120 miles a day. Camels can travel as fast as a horse and can work for long periods of time without food or water.

The feet of camels are wide and spread under their weight so as not to sink into the sand. Their long legs hold their bodies well above the hot sand, protecting them from heat radiation. Camels possess several attributes to protect them from blowing sand frequently encountered in the desert: (1) their nostrils can be closed; (2) dense eyelashes on both upper and lower eyelids; (3) a third transparent eyelid called the nictitating membrane; (4) a dense supply of hairs in the ear canal. When closed, this membrane allows the camel to see where it is going while protecting the eyes from blowing sand.

The camel's hump is used to store fat and allows the camel to go without food for several months. The camel can go without water for at least a week. When the fat store becomes depleted, a camel's hump becomes slack and hangs to one side. A myth persists that the camel stores water in its hump. There is no truth to this rumor. Camels do not store fat subcutaneously as humans do. This serves as heat insulation to keep humans warm. The camel can well do without such insulation during the day in the Sahara. The nights can be cold, but the camel is a large animal with hirsute body to keep it warm at night.

Camels live for up to fifty years. They usually have one calf at a time, though twins sometimes are born. About 90% of all camels are dromedaries. Camel products used by man include leather, milk, meat, wool, and even dung for fuel.

There are not many opportunities to drink in the desert, but when the camel does drink, it is able to swallow 13 gal (50 L) at one time. This would be fatal for any other animal due to electrolyte dilution, causing heart arrhythmia. Tissues would absorb so much water that the brain would swell, causing seizures and even death. This is probably the source of the myth that if you allow a camel to drink its fill, it will drink itself to death. This myth is untrue.

Camels are able to withstand body temperatures of 106°F (41°C) during the heat of the day. Camels can tolerate the loss of 30% of their body weight in water. Fluid levels in their blood can be replenished by borrowing water from other tissues. Their red blood cells (erythrocytes) resist water loss more than those of other animals. The camel's erythrocytes are oval not circular as in other mammals. As the blood thickens due to water loss, this oval shape helps blood to flow through capillaries. Normally camel blood contains 94% water, just as that of humans. The camel can safely lose up to 40% of its water, whereas a loss of 10% in humans results in insanity and loss of hearing. A loss of 12% water is fatal for humans.

Camels chew their cud like cows do. Camels are said to spit at people. There is some truth to this, but unfortunately, it is not saliva, but rather their cud that is being thrown at you. The camel regurgitates partially digested food and, with a shake of the head, throws it on the victim. Llamas, close relatives of camels, behave in a similar way, but they may spray saliva. They also regurgitate food and spray this at you. This behavior is believed to be caused by a perceived threat.

Camels chew their cud (ruminate) like cows, but camels have a three-chambered stomach whereas cow stomachs have four chambers. Cows are called ruminants; camels are said to be pseudoruminants. (Personally, I think this is unfair. They should get full-ruminant status.) Many desert plants are thorny. There is not much vegetation in the desert, and camels have developed thick skin on lips and mouth for eating thorn-covered plants. This adaptation is also seen in elephants and giraffes.

The ratel is also known as the honey badger. Its scientific name is *Mellivora capensis*. Melli—is Latin for "honey," vora means "to eat," as in "carnivore." Cap—refers to the Cape of Good Hope of Africa and—ensis means "of or belonging to." Ratel is the Afrikaans (not a misspelling) word for the creature. The ratel has markings reminiscent of the common skunk to which it is

Ratel (honey badger)

related. Both are in the weasel or Mustelidae family. Like the common skunk, the ratel has a broad white stripe on its back from head to tail. Also like the skunk, its anal glands produce an intolerably odiferous secretion. A broad white stripe on the back is contrary to common color schemes for animals. Most animals have darker dorsal coloration and a lighter ventral surface, a color scheme which is more in keeping with camouflage when viewed either from above or below. The ratel's front feet are equipped with long very strong claws. Ratels range in length from 24 to 40 in. (60-102 cm) not including the tail, which is 6.3-12 in. (16-30 cm) long. Females weigh 5-10 lb. (2.3-4.5 kg) while males weigh from 20 to 30 lb. (9-14 kg).

Ratels are fierce carnivores, eating worms, termites, scorpions, porcupines, hares, and snakes. Among snakes it eats are pythons up to 9.8 ft. (3 m), venomous adders, cobras, and black mambas. They eat crocodiles up to 3.3 ft. (1 m). They can devour a 5 ft. (1.5 m) snake in fifteen minutes. The 2002 Guinness Book of Records dubbed the ratel as "the most fearless animal in the world." In a National Geographic documentary, an incident was reported in which a ratel stole a meal out of the mouth of a puff adder. The ratel ate the meal in front of the snake, then turned on the snake and began eating the adder. Adders are a favorite snake preyed upon by ratels. The ratel collapsed, having been bitten by the snake. After about two hours, the ratel recovered, finished eating the adder, and went on his way.

The ratel has such thick hide with a heavy layer of subcutaneous fat that the bite of a wild dog does not faze it. Porcupine quills, bee stings, and even snakebites rarely penetrate its hide. If seized by the back of the neck, it can turn inside its skin and inflict a bite on its tormentor. The ratel has been known to charge out of its burrow to severely wound horses, antelope, cattle, and even water buffalo.

Ratels raid beehives for the honey they crave. Ratels are often snared by beekeepers and, while snared, may be stung to death by the bees. The greater honeyguide bird is said to vocally signal the ratel to follow the bird to a beehive. After eating his fill, the bird is left to consume dead bees and bee larvae along with some honeycomb. Bacteria in the bird's gut digest the honeycomb for the bird. This is the only known example of an animal digesting honeycomb. The relationship between the greater honeyguide bird (*Indicator indicator*) and the ratel is often called symbiosis, though a more accurate term is mutualism. The honeyguide bird has been known to guide humans to bee hives, but guiding of ratels to hives has been questioned since no biologist has ever confirmed this behavior. There is an exhibit in the Smithsonian Museum in Washingotn celebrating the honeyguide-ratel interraction.

The sea otter is another member of the weasel (Mustelidae) family like the ratel. The sea otter (*Enhydra lutris*) is the largest member of the Mustelidae family, but one of the smallest of marine mammals (walruses, sea lions, and seals). Male sea otters weigh 48-99 lb. (22-45 kg). Females weigh 30-73 lb. (14-33 kg). In the scientific name, *Enhydra lutris,* the en and hydra are Greek for "in" and "water." The lutris is Latin for "otter." Unlike many other Mustelidae, the sea otters do not have strong scent glands such as are present in skunks and the ratel. Such scent glands would not be very effective in water. The sea otter ventures out of the water to rest on rocky shores, but spends nearly all of its time in the ocean.

Two sea otters holding hands to prevent drifting apart

Their forefeet are equipped with tough pads and retractable claws for griping slippery prey. Their back feet are large and webbed for propulsion through the water. Sea otters have a unique adaptation; their outermost toes are the longest. This longest toe would correspond to the little toe in a human. This serves them well in the water, but their progress on land is clumsy. Both the small ears and the nostrils of the otter can be closed when diving underwater. The sea otter is the only marine creature that can move boulders on the seafloor in search of food.

Sea otters live in cold-water climates but have no blubber for thermal insulation. They depend on dense fur to reduce heat loss to the cold ocean water they inhabit. Sea otters have no molting period and so have a dense fur coat year round. They groom themselves meticulously to get rid of continuously shedding hair. They also keep the undercoating fine fur blown full of air which remains trapped in their coat while in water. Because air is trapped in their fine fur, the sea otter's skin stays dry while submerged, minimizing heat loss. The sea otter has long water-resistant guard hairs with dense fine hairs next to its skin. The fur of the sea otter is the densest of all animals, nearly a million hairs per square inch. The density of their fur led to their being hunted nearly to extinction before governments woke up and protected them. In spite of their unparalleled fur insulation, sea otters must maintain a high rate of metabolism, and therefore heat production, to compensate for heat loss. Their metabolic rate is two or three times that of a land animal of similar size. They must also eat 25-38% of their body weight each day.

Sea otters are thoroughly adapted to life in the ocean, even mating and giving birth in the water, unlike other marine mammals such as fur seals. Sea otters can find food with their front paws and, like a northern fur seal or house cat, with their long whiskers. Sea otters are able to drink saltwater because their kidneys produce highly concentrated urine to rid the blood of salt. Camels and kangaroo rats excrete concentrated urine in order to conserve water. Penguins can drink from the ocean by excreting concentrated salt solution in their tears.

The air is permanently trapped in the sea otter's fur. This trapped air, together with its large lung capacity, 2.5 times that of a mammal of similar size makes the sea otter quite buoyant. Sea otters normally rest and sleep on the water's surface by floating on their backs. A pair of sea otters, sleeping on the ocean's surface, interlocks the front paws nearest each other to keep from drifting apart (see illustration). This pair is likely to be mother and young. A mating pair bonds only while the female is in estrus.

Sea otter mothers are incredibly devoted to their young. Sea otter mothers have been known to carry a baby otter for days after the baby has died. Males do not participate in raising young. When a mother sea otter leaves her baby to search for food, she will often wrap her baby in kelp in order to provide an anchor so the baby cannot drift away. Adult sea otters may wrap themselves in kelp when sleeping for the same reason. Baby otters are precocial; that is, their eyes are open at birth, and ten teeth can be seen.

Sea otters are one of the few mammals that are able to use tools. Sea otters have a loose fold of skin under their arms and across their chest. They use this fold to carry food and a stone from the seabed. The otter, floating on its back, places the stone on its chest and uses the stone as an anvil to smash thick shells in order to eat the soft tissues inside. The sea otters also use the stones to hammer abalone shells in order to dislodge them from the rocks on the seabed. Sea otters eat abalone, sea urchins, mollusks, crustaceans, and certain types of fish. Sea otters catch fish using their front paws, the only animal to do this, with the possible exception of river otters.

Sea otters may live twenty years in captivity, but ten to fifteen years in the wild. Earlier death in the wild may be due to worn-down teeth. Sea otters normally remain within a territory near the coast and within only a few kilometers along the coastline. Sea otters can survive amidst drift ice, but not in land-fast ice.

Sea otters are regarded as a "keystone species," a species which benefits the ecosystem in a manner disproportionate to their numbers. Sea otters keep the populations of sea urchins in check. Sea urchins eat kelp roots which releases the kelp from the sea floor, killing it. Kelp does not have a true root which gathers nutrients from the soil, but only acts as an attachment. Kelp grows in dense communities called kelp forests. The kelp is the source of life, and when it is gone, the result is called an urchin barren.

Archeologists have found evidence that sea otters were hunted by indigenous people for thousands of years. The Aleuts used powdered baculum[42] to cure fevers. Sea otters became known globally with explorations by Vitus Bering in 1740. He was sent by Russia to map the Arctic coast and to find passage from Siberia to North America. Bering and some of his crew perished in a shipwreck off Bering Island. For one hundred years following, the "Great Hunt" took place in which fur traders from around the world reduced populations of sea otters to a small fraction of their original numbers.

Natural predators of sea otter pups include sea lions and bald eagles. Orcas prey on adult sea otters. Off Big Sur in California, dead sea otters have washed up after suffering bites inflicted by great white sharks. Great white sharks are not known to actually eat sea otters, but their bites are fatal.

The Exxon Valdez oil spill in March of 1989 killed thousands of sea otters in Prince William Sound. The pelt of the sea otter repels water but does not repel oil. The oil soaks through to the otter's skin and renders the fur a good agent for heat transfer from the otter's body to the cold seawater. The otter dies of hypothermia. If the otter is still alive, it will try to groom itself by licking off the oil. This will bring about kidney and liver failure. The lungs are also damaged by inhaling fumes from the oil.

Aside from oil spills, many populations of sea otters have bounced back from results of the Great Hunt. Reintroduction programs have also met with some success. Numbers are lower than before the Great Hunt but are much improved from the nadir following that hunt. International legislation has been extremely beneficial though some poaching takes place and fishery workers occasionally kill sea otters because of competition for their catch.

Walrus pod on ice floe

A baculum is a bone in the penis of the sea otter as well as in many other animals. Bacula, for curiosities and as carvings for knife handles, can be purchased on the Internet. Some animal sources are whales, sea lions, walrus, and raccoons. In the Southern United States, it was fashionable, at one time, for a Southern lady to carry a raccoon baculum in her purse for use as a toothpick. These bacula were available in some Southern restaurants.

The walrus is related to seals, sea lions, and elephant seals. All of these large mammalian sea creatures are pinnipeds. Pinniped means "winged foot." There are three families of pinnipeds:

1. Odobenidae. Literally, this means "tooth walkers." The only species in this family is the walrus, and its most striking characteristic is a pair of tusks which in the male may be 3 ft. (0.9 m) long. Females also have somewhat shorter tusks. These tusks are the upper canine teeth. The very heavy walrus uses these tusks to dig into an ice floe to help drag him/her out of the water and onto the ice.
2. Otariidae or "eared seals," which includes northern fur seals and sea lions. These have small external ears.
3. Phocidae[43] or "true seals" have no external ears. A lack of external ears is a characteristic shared with walruses, but Phocidae do not have tusks like walruses.

A walrus swimming under an ice floe uses its tusks to break through the ice. The walrus may also break the ice with its head. Mother walruses are known to rescue their pups that have become lodged in cracks in the ice by breaking the ice to pieces with their tusks. A cross section of a walrus tusk reveals growth rings similar to tree rings. By counting these rings, one can estimate the age of the walrus. A Pacific bull walrus weighs 1,760-3,750 lb. (800-1,700 kg). His tusks are 3 ft. (0.9 m) long. Pacific cows are somewhat smaller and have 2 ft. (0.6 m) tusks. An Atlantic species is somewhat smaller. Walrus have been known to live as long as forty years, but most are probably killed by hunters before reaching their maximum life span. Thick pads on the upper lips and cheeks of the walrus are covered with vibrissae (sensitive heavy whiskers).

Walrus have short, broad forepaws with short claws. Their hind flippers are larger, and the outer digits are longer than those in the middle are. The digits on the rear flippers end in short claws. The rear flippers can be folded under, enabling them to walk on land or ice like the eared seals and unlike the true seals. The walrus is sparsely covered with hair, which becomes thinner with age. When walrus lie out on the land or ice, their skin may become pink. Blood flows to the surface in order to warm the animals cooled by the arctic air. The blood at the skin surface gives the walrus a pink glow. When the walrus is in the arctic water, muscles

---

[43]   The word Phocidae is reminiscent of the term phocomelia, a congenital abnormality in which the limbs are shortened like flippers. This condition occurred in infants born to mothers who took thalidomide for morning sickness. Thalidomide was used in Europe, but was never approved for use in the United States.

controlling the flow of blood constrict to prevent loss of heat to the water. This causes them to appear almost white while swimming. The walrus is protected from the cold by 3 in. (7.6 cm) of blubber beneath its skin. Walrus are highly social, and groups of a thousand may congregate on ice shelves or on rocky shores.

Except for young bulls, walrus are not normally aggressive toward man. Walrus will come to the aid of the wounded of their kind, often pushing them off an ice floe in order that they may escape from hunters. Disputes over access to females for mating are usually settled by vocal warnings and visual displays of their tusks. A smaller male will withdraw quickly without any serious challenge. On occasion, well-matched senior walruses will engage in fights involving tusks which can be brutal.

Walrus mate in January or February and give birth fifteen months later. The fetus grows for only eleven months of the fifteen-month gestation period. The embryo is dormant due to delayed implantation for several months. All pinnipeds undergo delayed implantation of the embryo for reasons not well understood. Calves are dependent on their mothers for a year and a half to two years and a half. Walrus are noisy animals and produce loud complex songs for mating and social communication.

The only natural predators of the walrus are polar bears and orcas. Indigenous people hunt walrus, killing several thousand a year. Walrus sometimes prey on northern fur seals, but principally eat clams, oysters, and worms. The walrus finds food on the seafloor in shallow waters off the continental shelf. Cold water favors growth of clams and oysters by nourishment from algae that grow on the surface. In cold water, as opposed to tropical water, the algae sink to the bottom, providing food for bivalves. The walrus sucks the soft tissues out of the shells. Walrus eat prodigious quantities of bivalves, perhaps seven thousand in a day. Walrus find food on the benthic (bottom) surface by sucking in water and blowing it against the mud in jets. They also use the vibrissae to locate shellfish.

There are just two species of orangutans, one living in Borneo and the other in Sumatra. The scientific name of the Bornean species is *Pongo pygmaeus*, and the Sumatran species is *P. abelii*. *P. pygmaeus* is an endangered species, and *P. abelii* is critically endangered. Orangutans spend nearly all of their time in trees. Orangutans are reddish brown. The Sumatran species is of a lighter

Orangutan mother feeding her baby

shade and more sparsely covered than the Bornean. Males may weigh 250 lb. (110 kg). Females are smaller. Large males may have an arm span of 7.5 ft. (2 m). Both their hands and feet possess opposable digits, and they can grasp objects with either hands or feet. They are capable of delicate manipulations of small objects with their hands. Their arms are longer than their legs. The forearms are longer than the upper arms. Adult males have prominent cheek flaps which grow larger with age. Orangutans are threatened by habitat loss and by the sale of baby orangutans as pets. When taken from its mother, it is common for the baby eventually to die.

Orangutans are omnivorous, but their favorite food is fruit. Orangutans are the only agent of dispersal of the seeds of the climbing vine, *Strychnos ignatii*, which is a source of the toxic alkaloid strychnine. Aside from excessive salivation, eating the fruit of this plant seems to have no deleterious effect on orangutans.

A few experiments have been carried out to see if orangutans can learn sign language. Extensive work has shown that chimpanzees have learned signs remarkably well. Limited studies have resulted in orangutans learning thirty or forty signs.

The following two stories about the behavior of orangutans in zoos were related in Eugene Linden's book, *The Parrot's Lament* (see Suggested Readings). At the National Zoo in Washington, Rob Shumaker related this story about Bonnie, a female orangutan. Bonnie was holding her sick baby, Kiko. The keepers needed to give the baby orangutan an injection to fight an infection. They didn't want to tranquilize the mother as this would be traumatic for both mother and baby. Rob asked the veterinarian if he might try something. He approached the cage, and removing the syringe from his pocket, he showed it to Bonnie. He said, "Bonnie, I need to give the baby this." Bonnie had been administered injections before and knew that they were painful. But trusting that Rob meant well, she overcame her maternal instincts and brought the baby forward to the bars. She turned the baby around to receive the injection. Though the baby screamed in protest, Bonnie held it still until Rob finished. Rob and the vet were amazed at the understanding and trust shown by the mother orangutan.

Zoo designers use orangutans to test enclosures for possible avenues of escape. When an enclosure planned for a chimpanzee exhibit was built at the Los Angeles Zoo, officials brought in an orangutan to see if he could escape. Orangutans can be relied upon to attempt to escape. They reasoned that if an orangutan failed to break out, the enclosure would safely hold chimpanzees. Orangutans have more strength, a longer attention span, and perhaps more motivation to escape than chimpanzees do. Interestingly, most zoo animals including orangutans, chimpanzees, and even birds return voluntarily to their enclosures after they escape.

An orangutan named Fu Manchu was in a zoo in Brownsville, Texas. During the off-season when the zoo was closed to the public, the orangutans were allowed to occupy an outdoor exhibit, weather permitting. On such a day, a keeper came running to Jerry Stones, the zoo's curator. He told Stones that the orangutans were out and in the trees. A door led from the moat into the furnace room and then to a stairway to the outside. Stones was dismayed, but much more so when the same thing happened again. As a third attempt was in progress, one of the keepers was at hand to see what was happening. He quickly ran to Stones and told him that he had to see what Fu Manchu was up to. Positioning themselves on a nearby hillock, they watched Fu Manchu descend into the moat. He grabbed the bottom of the steel door and pulled it a little away from its casing. Orangutans are incredibly strong. Then he used a stiff wire to pop the door lock. He and the rest of the orangutans proceeded to freedom one more time.

The next day as Stones was leading the orangutans from the outside exhibit back into their cages, he saw something protruding a little from Fu Manchu's lip. Stones pulled back the ape's lip and found the wire he was using to pop the door latch. Fu Manchu had bent the wire into a shape that allowed him to hold it under his lip comfortably until the opportunity arose when he could put it to use. The Associated Locksmiths of America made Fu Manchu an honorary member.

Cetaceans are aquatic mammals with paddlelike forelimbs (flippers). They breathe air through single or double blowholes on top of their heads. Generally, the body of a cetacean is smooth and hairless, though some have short hairs (vibrissae) on their heads. Most whales spend at least part of the year in cold climates; a few live exclusively in cold climates. None of the cetaceans are protected by fur like seals are. Cetaceans rely on thick layers of blubber for thermal insulation. There are about eighty species of cetaceans, which include various types of whales, dolphins, and porpoises. Toothed whales have a single blowhole and include dolphins, beluga, narwhal, and sperm whales. Another group of whales are the baleen whales, which feed by straining water through baleen plates hanging from either side of their upper jaw. These plates may be equipped with dense hair to aid filtration whereby tiny plankton are trapped as nourishment for the whale. Baleen whales have two blowholes (nostrils). Baleens include gray, blue, bowhead, humpback, and several kinds of minke whales. The term whale is not a technical term. It may be loosely defined as a large cetacean. The killer whale, or orca, for example, is the largest member of the dolphin family.

The narwhal (occasionally also spelled narwhale) is a toothed whale, whose most notable feature is a 9.8 ft. (3 m) tusk projecting from the left side of the whale's upper jaw[44]. This tusk is a specialized canine tooth and occurs only in the male. The tusk is spiral in shape with a left thread as viewed from its base. Very rarely there may be tusks on both right and left sides, and they may be of about the same length. A smaller tusk may appear in the female, but this is rare. Narwhals are mammals and nurse their young.

Traders and chemists did their best to promote the belief in the existence of unicorns, as sale of narwhal tusks was extremely profitable as long as they were believed to have come from unicorns. Chemists were involved with another tall tale to the effect that goblets fashioned from unicorn's horn had the power to neutralize poisons. These were popular items with the nobility who were a bit paranoid about being poisoned. Queen Elizabeth possessed a jewel-encrusted unicorn goblet.

The narwhal is one of two whales in the Monodontidae family; the other is the beluga whale. Monodont means "one-toothed." Other than the males' tusk, narwhals have no functional teeth. The beluga whale (another monodont) has eight or nine pairs of peglike teeth in both upper and lower jaws. Monodontidae is somewhat of a misnomer for the family. The

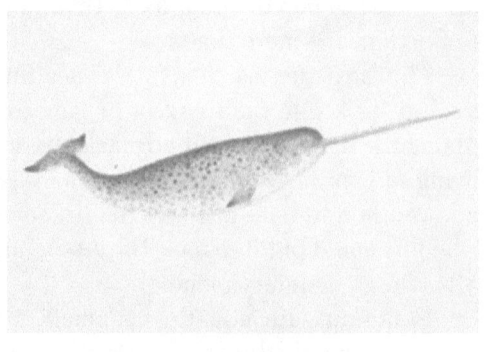

Narwhal

name narwhal comes from the Danish nar, meaning "corpse." The dappled gray and white of the adult whale was reminiscent of a drowned sailor.

Narwhals live in arctic waters near Greenland and Iceland in the Atlantic and also in parts of the northern Barents Sea. They live year round in arctic waters[45]. During the summer, they tend to be social, gathering in groups of about twenty, which tend to be similar in sex or age. These groups associate with others that together may number several hundred. In the winter, they tend to be more solitary, probably because narwhals stay close to pack ice

---

[44] One cannot help but be reminded of the fabled unicorn. In modern times, we picture a unicorn as a horse, often white, with a horn in the middle of his head. In medieval times, the unicorn tradition was well established, and many truly believed that the creature existed. The unicorn does not occur in Greek mythology, but rather in writings on Greek natural history. Leonardo da Vinci wrote about a scheme for capturing a unicorn by presenting it to a seated virgin. The unicorn would lay its head on her lap and go to sleep and could then be taken by hunters. The unicorn represented Christ in religious art.

[45] Narwhals, belugas, and the bowhead are the only whales that live year round in the Arctic. The bowhead is a robust whale with the head taking up one-third of its body length. It is said that the bowhead can burst through ice 6 ft. (1.8 m) thick.

where there is less open water. They may find themselves in a patch of water surrounded by ice. In this situation, they are vulnerable to polar bears. Orcas also prey upon narwhals.

Narwhals mate in spring, around April. Gestation lasts fifteen months, and most calves are born in July or August. The mother nurses her calf for at least a year. The calving interval is, perhaps, two, and more likely, three years.

Narwhals make deep dives in order to feed, reaching depths of 3,300 ft (1,000 m). They forage at all depths, taking arctic cod, squid, shrimp, and bottom dwellers such as Greenland halibut. They appear to suck prey into their mouths and swallow them whole. Sometimes, rocks are accidentally ingested in this process.

The orca, or killer whale, is a toothed whale (Odontoceti). Whale is not a technical term and is used to designate any large cetacean. Orca (*Orcinus orca*) is the largest of all dolphins. Its marking is strikingly black and white (see figure). Males may be 23 ft (9.6 m) long and weigh as much as 4 tons. Females are smaller. The fluke (tail) is black on the dorsal side and white on the ventral surface. The dorsal fin of the male can reach a length of 6 ft (1.8 m) with straight edges and is triangular. The female dorsal fin may be 3 ft high (0.9 m) and is falcate (curved) on the caudal side.

Two orcas breaching

The teeth of the orca are sharp, conical, and interlocking from above and below. Orcas are social and often travel in pods. Orcas are carnivorous and occur in all oceans. They are one of the few cetaceans that readily travel back and forth between northern and southern hemispheres. Three types of pods include offshore, transient, and residents. Their diet is dependent on the areas in which they forage. Transients consume a wide variety of animals such as emperor penguins, seals, sea lions, porpoises, squid, sharks, fish, and smaller whales.

Orcas may treat their prey with violence. Often, pups of pinnipeds (seals, sea lions, etc.) are thrown high into the air. As they land, the orcas bat them into the air again with their fluke. They thrash them against the water from side to side and finally swallow them. This treatment may result in skinning the pinniped pup. Orcas sometimes hunt in packs like wolves, earning them the sobriquet "wolves of the sea." Gray whales calve off Baja, Mexico. When the calves are old enough, the grays migrate to cold waters off Alaska to feed upon the great abundance of krill. The orca pod harasses the calf, which is unable to swim as fast as its mother can. Eventually the calf tires and the orcas are able to insert themselves between mother and calf. They then pile on top of the calf, covering its blowhole and forcing the calf underwater until it drowns.

Orcas have no natural enemies, other than man. Orcas have not been hunted as intensively as some other whales. Their numbers appear to be fairly stable. Overfishing may present a challenge to orcas since orcas eat fish. Pinnipeds also depend on fish, and pinnipeds are a major source of food for transient orcas.

Resident pods, which tend to remain in a restricted locale, tend to be of stable composition. They may consist of mothers and their calves, which remain with their mother throughout their life. Transient pods are more fluid, individuals coming and going.

Orcas seem to have a fond and cooperative relationship with humans as is also clearly seen with other smaller dolphins. There is no record of orcas attacking a human in the wild. Orcas have attacked their trainers in theme parks, sometimes with fatal consequences. Before Europeans arrived in Australia (1788), aborigines employed orcas to round up and kill whales. Later, orcas became partners with Europeans in the whaling industry. There was a pod of resident orcas in Twofold Bay which led whalers in the pursuit of whales. When the pod spotted a whale, they would surround it and dispatch messengers to the whaling village of Eden in Twofold Bay. The orca messengers would call the whalers by slapping the water with their flukes. The orcas would then guide the boats to the site of the besieged whale. With boats standing by, two orcas would grasp the whale's tail, preventing the whale from using it as a weapon. Two other orcas would swim under the whale to prevent it from submerging. Then other orcas would launch themselves onto the whale's blowhole, interfering with the whale's respiration. Finally, the exhausted whale would roll on the surface, and the whalers would drive their lances home. As the whale sunk to the bottom, the orcas fought to seize their prize, the whale's lips and tongue. This was the only part of the whale that was eaten by the orcas. After a few days, bacterial action produced gases which floated the carcass to the surface. The whalers would then harvest the commercially useful parts of the whale. In 1930, Old Tom, the dominant orca in the pod, washed up onto shore. After the death of Old Tom, the orca pod ceased cooperating with the whalers. The skeleton of Old Tom is the major exhibit in Eden's museum to the present day.

Suggested Readings

Attenborough, David. *Life in Cold Blood*. Princeton: Princeton University Press, 2008.

Attenborough, David. *The Life of Birds*. Princeton: Princeton University Press, 1998.

Attenborough, David. *The Life of Mammals*. Princeton: Princeton University Press, 2002.

Blumberg, Mark Samuel. Body Heat: Temperature and Life on Earth. Cambridge: Harvard University Press, 2002.

Burton, Robert. *Bird Behavior*. New York: Alfred A. Knopf, 1985.

Carwardine, Mark. *Whales, Dolphins, and Porpoises*. New York: Dorling Kindersley, 1995.

Ensminger, Peter A. *Life Under the Sun*. New Haven: Yale University Press, 2001.

Ford, Brian J. *The Secret Language of Life*. New York: Fromm International, 2000.

Foster, Russell G., and Leon Kreitzman. *Rhythms of Life*. New Haven: Yale University Press, 2004.

Geist, Valerius. *Moose: Behavior, Ecology, Conservation*. Stillwater: Voyageur Press, 1999.

Halliday, Tim, ed. *Animal Behavior*. Norman: University of Oklahoma Press, 1994.

Linden, Eugene. *The Parrot's Lament*. New York: Dutton, 1999.

McGowan, Christopher. *The Raptor and the Lamb: Predators and Prey in the Living World*. New York: Henry Holt, 1997.

Reeves, Randall R. et al. *Guide to Marine Mammals of the World*. New York: Alfred A. Knopf 2002.